Biology of Soil Science

Biology of Soil Science

Charlotte Jhonson

Oxford Book Company
Jaipur ▪ New Delhi

Published By

Oxford Book Company

Regd. Off. "Bony Residency", Gate No. 2
Opp. Tilak Public School, Vishweswariya Nagar,
Gopalpura Road, Jaipur - 302018, India
Ph: 0141 - 2761280, Telefax: 2761381
email - oxfordbookjpr@gmail.com
website - oxfordbookcompany.com

Delhi Office: 102, 1st Floor, Satyam House,
4327/3, Ansari Road, Darya Ganj,
New Delhi - 110002, India
Ph: 011-45652440

Published in India, 2025

ISBN : 9788189473969

Price : ₹ 2195/-

Typeset by Shivangi Computers, Jaipur

Printed and bound in India by Thomson Press Ltd.,
New Delhi

Preface

The study of soil science is the first step in the arduous task of prevention of erosion and rehabilitation of soil. Scientific studies have resulted in several advances in this area and Benchmarks have been laid down for the benefit of soil conservation authorities, government agencies and lay farmers. The present publication attempts to bring together the basic principles of soil science and the dynamics of erosion in all forms, for the benefit of the student as well as researcher and practising professional associated with conservation of natural resources. The text has been sell supported by prolific tables, diagrams pictures and illustrations. The publication will be of immense use to all those interested in this vital scientific area of earth studies.

The earth we mine, the land that produces food, fibre, and shelter, and yet few of us can even define soil, let alone explain how it works, what it does, or how it changes. This book contains the soil information, root growth and agricultural management, nitrate leaching management, podzols, paramos soils, water repellant soils, rare earth elements, and much more. It also presents hundreds of entries on tillage, irrigation, erosion control, minerals, ground water, and soil degradation. Further, the book outlines the agricultural, environmental, industrial, and cultural components that affect soil productivity and offers quick access to peer-reviewed articles on all branches of soil science—from mineralogy and physics to soil management, restoration and global warming and assesses the physical and hydrological properties of soil in natural and agricultural ecosystems.

Charlotte Jhonson

Preface

Contents

Chapter 1

Introduction

Soils are one of the Earth's essential natural resources, yet they are often taken for granted. Most people do not realise that soils are a living, breathing world supporting nearly all terrestrial life. Soils and the functions they play within an ecosystem vary greatly from one location to another as a result of many factors, including differences in climate, the animal and plant life living on them, soil's parent material, the position of the soil on the landscape, and the age of soil.

Scientists, engineers, farmers, developers and other professionals consider a soil's physical and chemical characteristics, moisture content and temperature to make decisions such as:

- Where is the best place to build a building?
- What types of crops will grow best in a particular field?
- Will the basement of a house flood when it rains?
- How can the quality of the groundwater in the area be improved?

Using the data collected in the Globe, Soil Investigation, students help scientist describe soils and understand how they function. They determine how soils change and the ways they affect other parts of the ecosystem, such as the climate, vegetation, and hydrology. Information about soils is integrated with data from the other Globe protocol investigations to gain a better view of Earth as a system.

INVESTIGATE SOILS

Soils develop on top of the Earth's land surface as a thin layer, known as the *pedosphere*. This thin layer is a precious natural resource and so deeply affects every part of the ecosystem that it is often called the "great integrator". For example, soils hold nutrients and water for plants and animals. They filter and clean water that passes through them. They can change the chemistry of water and the amount that recharges the groundwater or returns to the atmosphere to form rain.

The foods we eat and most of the materials we use for paper, buildings, and clothing are dependent on soils. Soils play an important role in the amount and types of gases in the atmosphere. They store and transfer heat, affect the temperature of the atmosphere, and control the activities of plants and other organisms living in soil. By studying these functions that soils play, students and scientists learn to interpret a site's climate, geology, vegetation, hydrology, and human history. They begin to understand soil as an important component of every ecosystem on Earth.

Scientists Need Globe Data

The Globe soil measurements are invaluable to scientists in many fields. For example, soil scientists use the data to better understand how soils form, how they should be managed, and what their potential is for plant growth and other land use. Hydrologists use the data to determine water movement through a soil and a watershed and the effect of soils on water chemistry. They also examine the effects of different types of soil on the sedimentation in rivers and lakes. Climatologists use soil data in climate prediction models. Atmospheric scientists want to know the effect of soils on humidity, temperature, reflected light, and fluxes of gases such as CO_2 and methane. Biologists examine the properties of soil to understand its potential for supporting plant and animal life.

Anthropologists study soil in order to reconstruct the human history of an area. When data are available for many areas of the world, scientists study the spatial patterns of soil properties. When a full set of GLOBE atmosphere, hydrology, land cover and soils data exists at a specific site, scientists can use the information to run computer models to understand how the whole ecosystem functions and to make predictions about what the ecosystem will be like in the future.

Composition of Soil

Soils are composed of four main components:

- Mineral particles of different sizes.
- Organic materials from the remains of dead plants and animals.
- Water that fills open pore spaces.
- Air that fills open pore spaces.

The use and function of a soil depends on the amount of each component. For example, a good soil for growing agricultural plants has about 45% minerals, 5% organic matter, 25% air, and 25% water. Plants that live in wetlands require more water and less air. Soils used as raw material for bricks need to be completely free of organic matter.

The Five Soil Forming Factors

The properties of a soil are the result of the interaction between the five soil forming factors. These factors are:

- *Parent Material:* The material from which soil is formed

determines many of its properties. The parent material of a soil may be bedrock, organic material, construction material, or loose soil material deposited by wind, water, glaciers, volcanoes, or moved down a slope by gravity.

- *Climate:* Heat, rain, ice, snow, wind, sunshine, and other environmental forces break down parent material, move loose soil material, determine the animals and plants able to survive at a location, and affect the rates of soil forming processes and the resulting soil properties.
- *Organisms:* Soil is home to large numbers of plants, animals, and microorganisms. The physical and chemical properties of a soil determine the type and number of organisms that can survive and thrive in that soil. Organisms also shape the soil they live in. For example, the growth of roots and the movement of animals and microorganisms shift materials and chemicals around in soil profile. The dead remains of soil organisms become organic matter that enriches the soil with carbon and nutrients. Animals and microorganisms living in the soil control the rates of decomposition for organic and waste materials. Organisms in soil contribute to the exchange of gases such as carbon dioxide, oxygen, and nitrogen between soil and the atmosphere. They also help soil filter impurities in water. Human actions transform soil as well, as we farm, build, dam, dig, process, transport, and dispose of waste.
- *Topography:* The location of a soil on a landscape also affects its formation and its resulting properties. For example, soils at the bottom of a hill will get more water than soils on the hillside, and soils on slopes that get direct sunlight will be drier than soils on slopes that do not.
- *Time:* The amount of time that the other 4 factors listed above have been interacting with each other affects the properties of the soil. Some properties, such as temperature and moisture content, change quickly, often over minutes and hours. Others, such as mineral changes, occur very slowly over hundreds or thousands of years.

THE SOIL PROFILES

The five soil forming factors differ from place to place causing soil properties to vary from one location to another. Each area of soil on a landscape has unique characteristics. A vertical section at one location is called a *soil profile.* When we look closely at the properties of a soil profile and consider the five soil forming factors, the story of soil at that site and the formation of the area is revealed.

These layers are known as *horizons.* Soil horizons can be as thin as a few millimetres or thicker than a metre. Individual horizons are identified by the properties they contain that are different from the horizons above and below them.

Table:1. Soil Properties That Change Over Time

Properties that change over Minutes or hours	Properties that change over months or years	Properties that change over hundreds and thousands of years
Temperature Moisture content Local composition of air	Soil pH Soil colour Soil structure Bulk density Soil organic matter Soil fertility Microorganisms, animals, Plants	Mineral content Particle size distribution Horizons Particle density

Some soil horizons are formed as a result of the weathering of minerals and decomposition of organic materials that move down the soil profile over time. This movement, called *illuviation,* influences the horizon's composition and properties. Other horizons may be formed by the disturbance of the soil profile from erosion, deposition, or biological activity. Soils may also have been altered by human activity. For example, builders compact soil, change its composition, move soil from one location to another, or replace horizons in a different order from their original formation.

Moisture in the Soil

Moisture plays a major role in the chemical, biological and physical activities that take place in soil. Chemically, moisture transports substances through the profile. This affects soil properties such as colour, texture, pH, and fertility. Biologically, moisture determines the types of plants that grow in soil and affects the way the roots are distributed. For example, in desert areas where soils are dry, plants such as cacti must store water, or send roots deep into soil to tap water buried tens of metres below the surface.

Plants in tropical regions have many of their roots near the surface where organic material stores much of the water and nutrients the plants need. Agricultural plants grow best in soils where water occupies approximately one-fourth of the soil volume as vapour or liquid.

Physically, soil moisture is part of the hydrologic cycle. Water falls on the soil surface as precipitation. This water seeps down into soil in a process called *infiltration.* After water infiltrates soil, it is stored in the horizons, taken up by plants, moved upward by *evaporation,* or moved downward into the underlying bedrock to become *ground water.* The amount of moisture contained in a soil can change rapidly, sometimes increasing within minutes or hours. In contrast, it might take weeks or months for soils to dry out.

If a soil horizon is compacted, has very small pore spaces, or is *saturated* with water, infiltration will occur slowly, increasing the potential for flooding in an area. If the water cannot move down into soil fast enough, it will flow over the surface as *runoff* and may rapidly end up in streams or other water bodies. When soil is not covered by vegetation and the slope of the land is steep, *water erosion* occurs. Deep scars are formed in the landscape as a result of the combined force of the runoff water and soil particles flowing over the surface. When a soil horizon is dry, or has large pore spaces that are similar in size to the horizon above, water will infiltrate the horizon quickly. If soil gets too dry and is not covered by vegetation, *wind erosion* may occur.

Soil Temperature

The temperature of a soil can change quickly. Near the surface, it changes almost as quickly as the air temperature changes, but because soil is denser than air, its temperature variations are less. Daily and annual cycles of soil temperature can be measured. During a typical day, soil is cool in the morning, warms during the afternoon, and then cools down again at night. Over the course of the year, soil warms up or cools down with the seasons.

Because soil temperature changes more slowly than air temperature, it acts as an insulator, protecting soil organisms and buried pipes from the extremes of air temperature variations. In temperate regions, the surface soil may freeze in winter and thaw in the spring, while in some colder climates, a permanent layer of ice, called *permafrost,* is found below the soil surface. In either case, the ground never freezes below a certain depth. The overlying soil acts as insulation so that the temperature of the deeper layers of soil is almost constant throughout the year. Temperature greatly affects the chemical and biological activity in soil. Generally, the warmer the soil, the greater the biological activity of microorganisms living in the soil.

Microorganisms in warm tropical soils break down organic materials much faster than microorganisms in cold climate soils. Near the surface, the temperature and moisture of soil affect the atmosphere as heat and water vapour are exchanged between the land and the air. These effects are smaller than those at the surfaces of oceans, seas, and large lakes, but can significantly influence local weather conditions. Hurricanes have been found to intensify when they pass over soil that is saturated with water. Meteorologists have found that their forecasts can be improved if they factor soil temperature and moisture into their calculations.

GLOBE MEASUREMENTS

In the Globe Soil Investigation, two sets of soil measurements are made. The first set, known as Soil Characterisation, describes the physical and chemical characteristics of each horizon in a soil profile. Some soil characterisation measurements are carried out in the field, while others are done in a laboratory or classroom. Soil characterisation measurements are carried out one time for an identified site. The second set of measurements are Soil Moisture and Temperature, which determine the water and temperature properties of soil at specified depths.

Soil moisture and temperature measurements are carried out repeatedly and can be directly compared with the air temperature and precipitation measurements that are described in the *Atmosphere Investigation*.

Although these 2 sets of soil measurements are different, having both soil characterisation and soil moisture at a given location provides the most amount of meaningful information. For example, differences in soil temperature and moisture between one site and another that have the same air temperature and precipitation may be due to differences in the soil characterisation properties. Understanding the physical and chemical properties of soil will help to interpret patterns in soil moisture and temperature.

Soil Characterisation Measurements Carried Out in the Field

- Site Description.
- Horizon Depths.
- Soil Structure.
- Soil Colour.
- Soil Consistence.
- Soil Texture.
- Roots.
- Rocks.
- Carbonates.

Carried Out in the Classroom or Lab

- Bulk Density.
- Particle Density.

- Particle Size Distribution.
- pH
- Soil Fertility (N, P, K).

Soil Moisture and Temperature Measurements Carried Out in the Field

- Soil Temperature.
- Soil Moisture Monitoring.
- Gravimetric Soil Moisture.

SOIL CHARACTERISATION

At a soil site, horizons in a soil profile are distinguished from one another by differences in their structure, colour, consistence, texture, and the amount of roots, rocks, and free carbonates they contain. Laboratory or classroom analyses of bulk density, particle density, particle size distribution, pH, and soil fertility also reveal differences among horizons.

Structure

Structure refers to the natural shape of aggregates of soil particles, called *peds*, in soil. The Soil structure provides information about the size and shape of pore spaces in soil through which water, heat, and air flow, and in which plant roots grow. Soil ped structure is described as *granular, blocky, prismatic, columnar,* or *platy*. If the soil lacks structure, it is described as either *single grained* or *massive*.

Colour

The Colour of soil is determined by the chemical coatings on soil particles, the amount of organic matter in the soil, and the moisture content of the soil. For example, soil colour tends to be darker when organic matter is present. Minerals, such as iron, can create shades of red and yellow on the surface of soil particles. Soil in dry areas may appear white due to coatings of calcium carbonate on the soil particles. Soil colour is also affected by moisture content. The amount of moisture contained in the soil depends on how long the soil has been freely draining or whether it is saturated with water. Typically, the greater the moisture content of a soil, the darker its colour.

Consistence

Consistence describes the firmness of the individual peds and the degree to which they break apart. The terms used to describe soil consistence are *loose, friable, firm,* and *extremely firm.* A soil with friable consistence will be easier for roots, shovels, or plows to move through than a soil with a firm consistence.

Texture

The Texture describes how a soil feels and is determined by the amounts of *sand, silt,* and *clay* particles present in the soil sample.

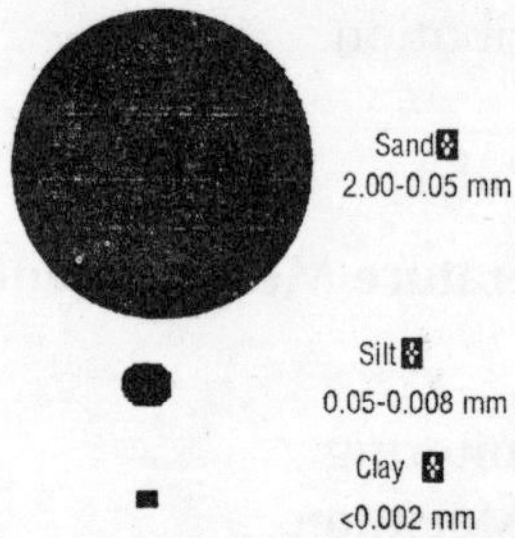

Fig:1. Particle Size Groups

The Soil texture influences how much water, heat, and nutrients will be stored in the soil profile. Human hands are sensitive to the difference in size of soil particles.

Sand is the largest particle size group, and feels gritty. Silt is the next particle size group, and feels smooth or *floury*. Clay is the smallest particle size group and feels sticky and is hard to squeeze.

Roots

The more roots found in a horizon, the more water and nutrients being removed from the soil, and the more organic matter being returned. Knowing the amount of roots in each horizon allows scientists to estimate the soil's fertility, bulk density, water holding capacity, and its depth. For example, a very compact horizon will inhibit root development whereas a porous horizon will not.

Rocks

An estimate of the number of rocks in each horizon helps to understand the movement of water, heat, and air through the soil, root growth, and the amount of soil material involved in chemical and physical reactions. Soil particles greater than 2 mm in size are considered to be rocks.

Carbonates

Carbonates of calcium or other elements accumulate in areas where there is little weathering from water. The presence of carbonates in soil may indicate a dry climate or a particular type of parent material rich in calcium, such as limestone. Free carbonates often coat soil particles in soils that are basic (i.e., pH greater than 7). These soils are common in arid or semi-arid climates.

Carbonates are usually white in colour and can be scratched easily with a fingernail. Sometimes in dry climates, carbonates form a hard and dense horizon similar to cement, and plant roots cannot grow through it. To test for carbonates, an acid, such as vinegar, is squirted on soil. If carbonates are present, there will be a chemical reaction between the vinegar (an acid) and the carbonates to produce carbon dioxide. When carbon dioxide is produced, the vinegar bubbles or *effervesces*. The more carbonates present, the more bubbles or effervescence occurs.

Bulk Density

Soil *bulk density* is a measure of how tightly packed or dense the soil is and is measured by the mass of dry soil in a unit of volume (g/cm^3). Soil bulk density depends on the composition of soil, structure of the soil peds, the distribution of the sand, silt, and clay particles, the volume of pore space, and how tightly the particles are packed. Soils made of minerals (sand, silt, and clay) will have a different bulk density than soils made of organic material. In general, the bulk density of soils ranges from 0.5 g/cm^3 in soils with many spaces, to as high as 2.0 g/cm^3 or greater in very compact mineral horizons.

Knowing the bulk density of a soil is important for many reasons. Bulk density indicates how tightly soil particles are packed and the ease with which roots can grow through soil horizons. Bulk density is also used when converting between mass and volume for a soil sample. If the mass of a soil sample is known, its volume is calculated by dividing the sample mass by the bulk density of the soil. If the volume of a soil sample is known, the mass is calculated by multiplying the sample volume by the bulk density of the soil.

Particle Density

The *Particle density* of a soil sample is the mass of dry soil in a particular volume of soil when all of the air spaces have been removed. The type of minerals the soil particles are made of affects the particle density. Soils consisting of pure quartz particles generally have a particle density of 2.65 g/cm^3.

Soils consisting of particles made of minerals other than quartz will have a different mass for the same volume of particles. By knowing both the particle density and the bulk density, the *porosity* can be calculated. Porosity establishes the amount of air or water that can be stored or moved through the soil.

Particle Size Distribution

The proportion of each particle size group (sand, silt, or clay) in the soil is called the soil *particle-size distribution.* Sand is the largest soil particle, silt is intermediate in size, and clay is the smallest. The particle-size distribution of a soil sample determines its exact textural class. It also helps determine how much water, heat, and nutrients soil will hold, how fast water and heat will move through soil, and the structure and consistence of soil.

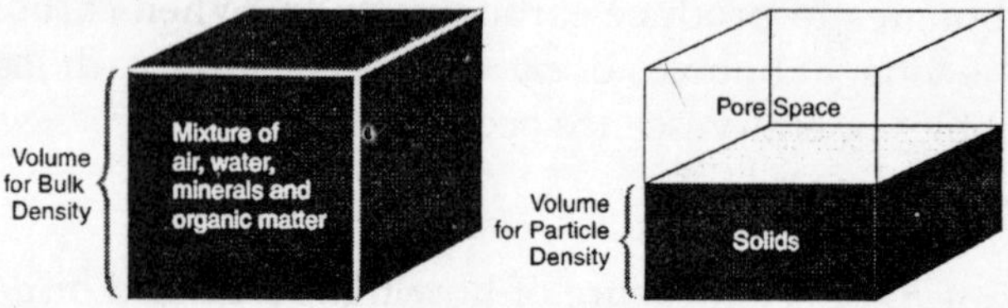

Fig:2. A Comparison of Bulk Density and Particle Density

The amount of sand, silt, and clay in a soil sample is determined by a settling method using an instrument called a *hydrometer*. A dried sample of soil is first dispersed so that none of the particles stick together, and then it is suspended in water and allowed to settle.

The largest particles (sand) settle out in minutes while the smallest particles (clay) stay suspended for days. A hydrometer is used to measure the specific gravity of the soil suspension after settling has proceeded for specific amounts of time.

The pH

The *pH* of a soil horizon is determined by the parent material from which soil is formed, the chemical nature of the rain or other water entering soil, land management practices, and the activities of organisms (plants, animals, and microorganisms) living in soil. Just like the pH of water, the pH of soil is measured on a logarithmic scale. The soil pH is an indication of the soil's chemistry and fertility.

The activity of the chemical substances in soil affects the pH levels. Different plants grow at different pH values. Farmers sometimes add materials to soil to change its pH depending on the types of plants they want to grow. The pH of soil also affects the pH of ground water or nearby water bodies such as streams or lakes. Soil pH can be related to the water pH measured in the *Hydrology Investigation* and the precipitation pH measured in the *Atmosphere Investigation*.

The Fertility

The *fertility* of a soil is determined by the amount of nutrients it contains. Nitrogen (N), phosphorus (P), and potassium (K) are three of the most important nutrients needed by plants for optimum plant growth. Each horizon in a soil profile can be tested for the presence of these nutrients. The results of these measurements help to determine the suitability of a soil for growing plants. The soil fertility can be related to water chemistry measurements carried out in the *Hydrology Investigation*.

SOIL MOISTURE

Soil moisture, also known as *Soil Water Content* (SWC), is a ratio of the mass of water contained in a soil sample to the mass of dry matter in that sample.

This ratio typically ranges from 0.05 g/g to 0.50 g/g. Only extremely dry soils that retain a small amount of water, such as those in a desert, have values below 0.05 g/g. Only organic-rich soils, peat or some clays absorb large amounts of water and have values above 0.50 g/g. In some very highly organic soils, the soil water content may actually be >1.0 g/g because the mass of the water is greater than the mass of the organic particles.

Soil moisture measurement helps to define the role of the soil in the surrounding ecosystem. For example, soil moisture measurements reveal the ability of the soil to hold or transmit water affecting groundwater recharge, surface runoff, and *transpiration* and evaporation of water into the atmosphere. It also describes the ability of the soil to provide nutrients and water to plants, affecting their growth and survival.

SOIL TEMPERATURE

Soil acts as an insulator for heat flowing between the solid earth below the soil and the atmosphere. Thus, soil temperatures can be relatively cool in the summer or relatively warm in the winter. These soil temperature variations affect plant growth, the timing of bud-burst or leaf fall, and the rate of decomposition of organic materials.

Soil temperatures typically have a smaller daily range than air temperatures and deeper soil temperatures usually vary less. Soil temperature extremes range from 50°C for near-surface summer desert soils to values below freezing in high latitude or high elevation soils in the winter.

SOIL STUDY SITE SELECTION

Soil study sites for carrying out soil characterisation measurements and soil moisture and temperature measurements should be carefully selected.

For soil characterization measurements, a site should be considered that allows students to dig a hole with either a shovel or an auger. The purpose is to expose a soil profile that is one metre deep. If this is not possible, students have the option to sample the top 10 cm of the soil profile. It is important to check with local utility companies to be sure there are no pipes or wires buried at the site chosen for digging. A site that is chosen close to the site where soil moisture and temperature measurements are being made will help to understand these measurements better. A soil characterisation site chosen near or in the Land Cover Study Site will help interpret the role that the soil properties play in controlling the type and amount of plant growth.

For soil moisture measurements, a site that is open should be considered. The site must not be irrigated, should have *uniform* soil characteristics, be relatively undisturbed, and be safe for digging. Soil moisture samples are collected from the surface (0-5 cm) and 10 cm depths. Samples may also be collected at depths of 30 cm, 60 cm, and 90 cm to obtain a depth profile. If possible, the site should be within 100 m of a Globe Atmosphere Study Site or other location where precipitation measurements are being collected.

For soil temperature measurements, a site should be selected that is adjacent to a Globe Atmosphere Study site, or some other location where air temperature measurements are taken. Alternatively, soil temperature can be measured at a soil moisture study Site. The site should be in the open and representative of soils in the area. Soil temperature measurements are made at depths of 5 and 10 cm with all protocols and also at 50 cm with monitoring protocols.

Site Description

After students have selected a site for their soil measurements, they use the following identifying factors to define and describe the location they plan to study: latitude and longitude (using GPS receivers), elevation, slope, aspect (the direction of the steepest slope), type of vegetation covering soil, parent material, current land use practices, and the position of soil on the landscape. The students determine some of these properties at the site, while other properties are established using local resources such as maps, soil survey reports, and local experts.

Frequency of Measurements

Soil characterisation measurements should be carried out one time for each Soil Characterisation Study Site. More than one study site can be used in order to identify soil properties at different locations (such as at the soil moisture and temperature sites, land cover site, or along different parts of the landscape for example).

To help in understanding the global picture of soil moisture, Globe's highest priority is soil moisture measurements carried out during two primary collection campaigns each spring and fall.

To study local changes, soil moisture measurements should be taken twelve or more times per year for the same site at weekly or monthly intervals. With soil moisture sensors, measurements should be taken daily or more frequently.

Soil temperature measurements are carried out at least once each week. Many schools take daily soil temperature measurements at the same time they collect daily atmospheric data. The *Digital Multi-Day Max/Min/ Current Air and Soil Temperature Protocol* provides for daily measurement of the maximum and minimum soil temperatures from a depth of 10 cm.

Optional protocols are available for measuring daily maximum and minimum soil temperatures at 5 cm and 50 cm depths and for collecting soil and air temperature automatically every 15 minutes using a data logger.

Combining the Measurements

In the Globe Soil Investigation, students study both the soil properties that change very slowly (soil characterization), and those that change rapidly (soil temperature and moisture). Without knowing the slowly changing properties of the soil profile, it is difficult to understand the dynamic moisture and temperature changes that occur. In the same way, the patterns in moisture and temperature in the soil over time, affect the formation of soil. Teachers are encouraged to combine soil characterisation measurements with soil temperature and moisture measurements so that students gain a true understanding of the way the pedosphere functions and affects the rest of the ecosystem.

SOIL

Soil the term taken from the latin word usage of *solea* means soil or ground, which in classic Latin meant the sole of the foot, also a sandal. This was due to a confusion with *solum,* ground, whence Fr. *sol.* Both *solea* and *solum* are, of course, from the same root. To be distinguished from this word is "soil", to make dirty, to stain, defile. The origin is the *soil* or *souil,* the miry wallowing ground of a wild boar, whence the hunting phrase "to take soil", of a beast of the chase taking to water or marshy ground. The derivation is therefore from Lat. *soillus,* pertaining to substance which is stirred or tilled by implements such as ploughs and spades. Below this is *subsoil.* Soil through being acted upon by the air, heat, frost and other agencies usually consists of finer particles than those. comprising the bulk of the subsoil. It contains more roots, and as a rule, is darker in colour than the subsoil on account of the larger proportion of decaying vegetable matter present in it; it is also looser in texture than the subsoil. Subsoil not unfrequently contains materials which are deleterious to the growth of crops, and roots descending into it may absorb and convey these poisonous substances to other parts of the plant or be themselves damaged by contact with them. On this account deeper tillage than usual, which allows of easier penetration of roots, or the carrying out of operations which bring subsoil to the surface, must always be carefully considered.

At first sight few natural materials appear to be of less interest than soil; yet its importance is manifest on the slightest reflection. From it, directly or indirectly, are obtained all food materials needed by man and beast. The inorganic materials within it supply some of the chief substances utilised by plants for their development and growth, and from plants animals obtain much of their sustenance.

Origin of the Soil

It is a matter of common observation that stones of monuments, walls or buildings which are exposed to the air sooner or later become eaten away or broken up into small fragments under the influence of the weather. This disintegration is brought about chiefly by changes in temperature, and by the action of the rain, the oxygen, and the carbon dioxide of the air. During the daytime the surface of the stone may become very warm, while at night it is speedily cooled. Such alterations in temperature produce strains which frequently result in the chipping off of small fragments of the material composing the stone.

Moreover, the rain penetrates into the small interstices between its particles and dissolves out some of the materials which bind the whole into a solid stone, the surface then becoming a loose powdery mass which falls to the ground below or is carried away by the wind. The action of frost is also very destructive to many stones, since the water within their cracks and crannies expands on freezing and splits off small pieces from their surfaces. In the case of limestones the carbon dioxide of the air in association with rain and dew eats into them and leads to their disintegration. The oxygen of the air may also bring about chemical changes which result in the production of soluble substances removable by rain, the insoluble parts being left in a loosened state.

These "weathering" agents not only act upon stones of buildings, but upon rocks of all kinds, reducing them sooner or later into a more or less fine powder. The work has been going on for ages, and the finely comminuted particles of rocks form the main bulk of soil which covers much of the earth's surface, the rest of the soil being composed chiefly of the remains of roots and other parts of plants.

If the whole of soil in the British Islands were swept into the sea and the rocks beneath it laid bare the surface of the country would ultimately become covered again with soil produced from the rocks by the weathering processes just described. Moreover, where there was no transport or solution of soil thus produced it would necessarily show some similarity in composition to the rock on which it rested. Soils overlying red sandstone rocks would be reddish and of a sandy nature, while those overlying chalk would be whitish and contain considerable amounts of lime. In many parts of the country soils exhibiting such relationships, and known as *sedentary* soils, are prevalent, the transition from soil to the rock beneath being plainly visible in sections exposed to view in railway cuttings, quarries and other excavations.

In many districts soil is manifestly unconnected in origin with the rock on which it rests, and differs from it in colour, composition and other characters. There are *transported* or drift soils, the particles of which have been brought from other areas and deposited over the rocks below. Some of the stiff boulder clays or " till " so prevalent over parts of the north of England appear to have been deposited from ice sheets during the glacial period. Perhaps the majority of drift soils, however, have been moved to their present position by the action of the water of rivers or the sea.

As fast as the rock of a cliff is weathered its fragments are washed to the ground by the rain, and carried down the slopes by small streams, ultimately finding their way into a river along which they are carried until the force of the water is insufficient to keep them in suspension, when they become deposited in the river bed or along its banks. Such river-transported material or *alluvium* is common in all river valleys. It is often of very mixed origin, being derived from the detritus of many kinds of rocks, and usually forms soil of a fertile character.

Quality of Soil

The good or bad qualities of a soil have reference to the needs of the crops which are to be grown upon it, and it is only after a consideration of the requirements of plants that a clear conception can be formed of what characters the soil must possess for it to be a suitable medium on which healthy crops can be raised. In the first place, soil, to be of any use, must be sufficiently loose and porous to allow the roots of plants to grow and extend freely. It may be so compact that root development is checked or stopped altogether, in which case the plant suffers. On the other hand, it should not be too open in texture or the roots do not get a proper hold of the ground and are easily disturbed by wind: moreover, such soils are liable to blow away, leaving the underground parts exposed to the air and drought.

The roots like all other parts of plants contain protoplasm or living material, which cannot carry on its functions unless it is supplied with an adequate amount of oxygen: hence the necessity for the continuous circulation of fresh air through the soil. If the latter is too compact or has its interstices filled with carbon dioxide gas or with water–as is the case when the ground is water-logged–the roots rapidly die of suffocation just as would an animal under the same conditions. There is another point which requires attention. Plants need very considerable amounts of water for their nutrition and growth; the waterholding capacity is, therefore, important. If the soil holds too much it becomes water-logged and its temperature falls below the point for healthy growth, at any rate of the kinds of plants. usually cultivated on farms and in gardens. If it allows too free drainage drought sets in and the plants, not getting enough water for their needs, become stunted in size. Too much water is bad, and too little is equally injurious.

In addition, the temperature of the soil largely controls the yield of crops which can be obtained from the land. Soil whose temperature remains low, whether from its northerly aspect or from its high water content or other cause, is unsatisfactory, because the germination of seeds and the general life processes of plants cannot go on satisfactorily except at certain temperatures well above freezing-point.

A good soil should be deep to allow of extensive root development and, in the case of arable soils, easy to work with implements. Even when all the conditions above mentioned in regard to texture, water-holding capacity, aeration and temperature are suitably fulfilled the soil may still be barren: plant food-material is needed. This is usually present in abundance although it may not be available to the plant under certain circumstances, or may need to be replenished or increased by additions to the soil of manures or fertilizers.

CHIEF CONSTITUENTS OF SOIL

An examination of soil shows it to be composed of a vast number of small particles of sand, clay, chalk and humus, in which are generally imbedded larger or smaller stones. It will be useful to consider the nature of the four chief constituents just mentioned and their bearing upon the texture.

Sand consists of grains of quartz or flint, the individual particles of which are large enough to be seen with the unaided eye or readily felt as gritty grains when rubbed between the finger and thumb. When a little soil is shaken up with water in a tumbler the sand particles rapidly fall to the bottom and form a layer which resembles ordinary sand of the seashore or river banks. Chemically pure sand is silicon dioxide (SiO_2) or quartz, a clear transparent glass-like mineral, but as ordinarily met with, it is more or less impure and generally coloured reddish or yellowish by oxide of iron. A soil consisting of sand entirely would be very loose, would have little capacity to retain water, would be liable to become very hot in the daytime and cool at night and would be quite unsuitable for growth of plants.

The term clay is often used by chemists to denote hydrated silicate of alumina ($Al_2O_3\ 2SiO_2.2H_2O$), of which kaolin or china clay is a fairly pure form. This substance is present in practically all soils but in comparatively small amounts. Even in soils which farmers speak of as stiff clays it is rarely present to the extent of more than 1 or 2%. The word "clay" used in the agricultural sense denotes a sticky intractable material which is found to consist of exceedingly fine particles (generally less than .005 mm. in diameter) of sand and other minerals derived from the decomposition of rocks, with a small amount of silicate of alumina. The peculiar character which clay possesses is probably due not to its chemical composition but to its physical state. When wet it becomes sticky and almost impossible to move or work with farm implements; neither air nor water can penetrate freely. In a dry state it becomes hard and bakes to a brick. It holds water well and is consequently cold, needing the application of much heat to raise its temperature. It is obvious, therefore, that soil composed entirely of clay is as useless as pure sand so far as the growth of crops upon it is concerned.

Chalk consists, when quite pure, of calcium carbonate ($CaCO_3$), a white solid substance useful in small amounts as a plant food-material, though in excess detrimental to growth. Alone, even when broken up into small pieces, it is unsuitable for the growth of plants.

Humus, the remaining constituent of soil, is the term used for the decaying vegetable and animal matter in the soil. A good illustration of it is peat. Its water-holding capacity is great, but it is often acid, and when dry it is light and incapable of supporting the roots of plants properly. Few of the commonly cultivated crops can live in a soil consisting mainly of humus.

A perfect soil would be such a blend of sand, clay, chalk and humus as would contain sufficient clay and humus to prevent drought, enough sand to render it pervious to fresh air and prevent waterlogging, chalk enough to correct the tendency to acidity of the humus present, and would have within it various substances which would serve as food-materials to the crops.

Generally speaking, soils containing from 30 to 50% of clay and 50 to 60% of sand with an adequate amount of vegetable residues prove the most useful for ordinary farm and garden crops; such blends are known as "loams", those in which clay predominates being termed clay loams, and those in which the sand predominates is termed sandy loams. "Stiff clays" contain over 50% of clay. In the mechanical analysis of soil, after separation of the stones and fine gravel by means of sieves, the remainder of the finer earth is subjected to various processes of sifting and deposition from water with a view of determining the relative proportions of sand, silt and clay present in it.

Most of the material termed "sand" in such analyses consists of particles ranging in diameter from 0.5 to 0.05 mm, and the "silt" from 0.05 to 0.005 mm, the "clay" being composed of particles less than 0.005 mm in diameter. The proportional amount of these materials in a sandy soil on the Bagshot beds. The pore-space within the soil, i.e. the space between the particles composing soil, varies with the size of these particles and with the way they are arranged or packed. It is important, since upon it largely depends the movement of air and water in the land. It is generally from 30 to 50% of the total volume occupied by soil. Where the soil grains are quite free from each other the smaller grains tend to fill up the spaces between the larger ones; hence it might be concluded that in clays the amount of pore-space would be less than in coarser sands. This is the case in "puddled" clays, but in ordinary clay soils the excessively minute particles of which they largely consist tend to form groups of comparatively large composite grains and it is in such natural soils that the pore-space is largest.

CHEMICAL COMPOSITION OF SOIL

It has been found by experiment that plants need for their nutritive process and their growth, certain chemical elements, namely, carbon, hydrogen, oxygen, nitrogen, sulphur, phosphorus, potassium, magnesium, calcium and iron. With the exception of the carbon and a small proportion of the oxygen and nitrogen, which may be partially derived from the air, these elements are taken from soil by crops. Plants also remove from the soil silicon, sodium, chlorine, and other elements which are, nevertheless, found to be unessential for the growth and may therefore be neglected here.

Leguminous crops take some of the nitrogen which they require from the air, but most plants obtain it from the nitrates present in soil. The sulphur exists in soil chiefly in the form of sulphates of magnesium, calcium and other metals; the phosphorus mainly as phosphates of calcium, magnesium and iron; the potash, soda and other bases as silicates and nitrates; calcium and magnesium carbonates are also common constituents of many soils.

In the ordinary chemical analyses of soil determinations are made of the nitrogen and various carbonates present as well as of the amount of phosphoric acid, potash, soda, magnesia and other components soluble in strong hydrochloric acid.

Below are given examples of the analyses of a poor sandy soil and an ordinary loam: Since the dry weight of the first foot of soil over an acre is about 4,000,000 lb the poor sandy soil contains within it: Nitrogen. 7,600 lb Phosphoric acid 7,200 Potash 7,600 Lime 9,200. Potash and phosphoric acid removed by a wheat or mangel crop it would appear that this soil has enough of these ingredients in it to yield many such crops; yet experience has shown that these crops cannot be grown on such a poor sandy soil unless manures containing phosphates, potash and nitrogen are added.

Many attempts have been made to correlate the results of the analyses of a soil with its known cropping power, but there is yet much to be learnt in regard to these matters. A great proportion of the food constituents which can be extracted by strong hydrochloric acid are not in a condition to be taken up by the roots of plants; they are present, but in a "dormant" state, although by tillage and weathering processes they may in time become "available" to plants. Analyses of this character would appear to indicate the permanent productive capacity of soil rather than its immediate power of growing a crop.

Soils containing less than 25% of potash are likely to need special application of potash fertilizers to give good results, while those containing as much as 4 or 5% do not usually respond to those manures. Where the amount of phosphoric acid is less than 05%, phosphatic manures are generally found to be beneficial with more than 1% present these fertilizers are not usually called for except perhaps in soils containing a high percentage of iron compounds.

Similarly soils with less than 1% of nitrogen are likely to be benefited by applications of nitrogenous manures. At present it is not possible to determine with accuracy the amount of immediately available plant food-constituents in a soil: no doubt the various species of plants differ somewhat in their power of absorbing these even from the same soil. The method introduced by Dyer of dissolving out the mineral constituents of the soil with a 1% solution of citric acid, which represents about the average acidity of the roots of most common plants, yields better results. In the case of arable soils, where the amount of phosphoric acid determined by this method falls below 1%, phosphatic manuring *is* essential for good crops.

The writer has found that many pasture soils containing less than 0.025 or 0.03%, respond freely to applications of phosphates; probably in such cases even the weak acid is capable of dissolving out phosphates from the humus or other compounds which yield little or none to the roots of grasses and clovers. In soils where the potash available to citric acid is less than 0.005%, kainit and other potash fertilizers are needed.

WATER IN SOIL

The importance of an adequate supply of water to growing crops cannot well be over-estimated. During the life of a plant there is a continuous stream of water passing through it which enters by the root-hairs in the soil and after passing along the stem is given off from the stomata of the leaves into the open air above ground. It has been estimated that an acre of cabbage will absorb from the land and transpire from its leaves more than ten tons of water per day when the weather is fine.

In addition to its usefulness in maintaining a turgid state of the young cells without which growth cannot proceed, water is itself a plant food-material and as absorbed from soil contains dissolved in it all the mineral food constituents needed by plants for healthy nutrition. Without a sufficient supply plants remain stunted and the crop yield is seriously reduced. If one condition is more necessary than another for good crops it is a suitable supply of water, for no amount of manuring or other treatment of soil will make up for a deficient rainfall. The amount needed for the most satisfactory nutrition varies with different plants. In the case of fair average farm crops it has been shown that for the production of one ton of dry matter contained in them from 300 to 500 tons of water has been absorbed and utilised by the plants. This may be more than the rainfall, in which case irrigation or special control of the water supply may be necessary.

The water-holding capacity of a soil depends upon the amount of free space between the particles of which it is composed into which water can enter. In most cases this amounts to from 30 to 50% of the volume of soil.

When the pore-space of soil is filled with water it becomes water-logged and few plants can effect absorption by their roots under such conditions. The root-hairs die from want of air, and the whole plant soon suffers. Fields of wheat and other cereals rarely recover after a week's submergence, but orchards and many trees when at rest in winter withstand a flooded or water-logged condition of the soil for two or three weeks without damage.

The most satisfactory growth is maintained when the amount of water present is not more than 40 to 60% of what would saturate it. Under such conditions each particle of soil is surrounded by a thin film of water and in the pore-space air can freely circulate. It is from such films that the root-hairs absorb all that plants require for their growth. The movement of water into the root-hairs is brought about by the osmotic action of certain salts in their cell-sap. Crops are, however, unable to absorb all the water present in soil, for when the films become very thin they are held more firmly or cling with more force to the soil particles and resist the osmotic action of the root-hairs. Plants have been found to wither and die in sandy soils containing 1% of water, and in clay soils in which there was still present 8% of water.

When a long glass tube open at both ends is filled with soil and one end is dipped in a shallow basin of water, the water is found to move upwards through soil column just as oil will rise in an ordinary lamp wick. By this capillary action water may be transferred to the upper layers of soil from a depth of several feet below the surface. In this manner plants whose roots descend but a little way in the ground are enabled to draw on deep supplies. Not only does water move upwards, but it is transferred by capillarity in all directions through soil. The amount and speed of movement of water by this means, and the distance to which it may be carried, depend largely upon the fineness of the particles composing soil and the spaces left between each. The ascent of water is most rapid through coarse sands, but the height to which it will rise is comparatively small.

In clays whose particles are exceedingly minute the water travels very slowly but may ultimately reach a height of many feet above the level of the "water-table" below. While this capillary movement of water is of great importance in supplying the needs of plants it has its disadvantages, since water may be transferred to the surface of soil, where it evaporates into the air and is lost to the land or the crop growing upon it. The loss in this manner was found to be in one instance over a pound of water per day per square foot of surface, the "water-table" being about 4 or 5 ft below.

One of the most effective means of conserving soil moisture is by "mulching", i.e. by covering the surface of the soil with some loosely compacted material such as straw, leaf-refuse or stable manure. The space between the parts of such substances is too large to admit of capillary action; hence the water conveyed to the surface of soil is prevented from passing upwards any further except by slow evaporation through the mulching layer.

A loose layer of earth spread over the surface of soil acts in the same way, and a similarly effective mulch may be prepared by hoeing the soil, or stirring it to a depth of one or two inches with harrows or other implements. The hoe and harrow are therefore excellent tools for use in dry weather. Rolling the land is beneficial to young crops. in dry weather, since it promotes capillary action by reducing soil spaces. It should, however, be followed by a light hoeing or harrowing.

So far as the water-supply is concerned - and this is what ultimately determines the yield of crops – the rain which falls upon the soil should be made to enter it and percolate rapidly through its interstices. A deep porous bed in the upper layers is essential, and this should consist of fine particles which lie close to each other without any tendency to stick together and "puddle" after heavy showers. Every effort should be made to prepare a good mealy tilth by suitable ploughing, harrowing and consolidation.

In the operation of ploughing the furrow slice is separated from the soil below, and although in humid soils this layer may be left to settle by degrees, in semi-arid regions this loosened layer becomes. dry if left alone even for a few hours and valuable water evaporates into the air.

To prevent this various implements, such as disk harrows and specially constructed rollers, may be used to consolidate the upper stirred portion of soil and place it in close capillary relationship with the lower unmoved layer. If the soil is allowed to become dry and pulverised, rain is likely to run off or puddle the surface without penetrating it more than a very short distance. Constant hoeing or harrowing to maintain a natural soil mulch layer of 2 or 3 in. deep greatly conserves the soil water below. In certain districts where the rainfall is low a crop can only be obtained once every alternate year, the intervening season being devoted to tillage with a view of getting the rain into soil and retaining it there for the crop in the following year.

Bacteria in Soil

Recent science has made much progress in the investigation of the microorganisms of soil. Whereas soil used to be looked upon solely as a dead, inert material containing certain chemical substances which serve as food constituents. of the crops grown upon it, it is now known to be a place of habitation for myriads of minute living organisms upon whose activity much of its fertility depends. They are responsible for many important chemical processes which make the soil constituents more available and better adapted to the nutrition of crops. One cubic centimetre of soil taken within a foot or so from the surface contains from 1 to 2 millions of bacteria of many different kinds, as well as large numbers of fungi. In the lower depths of soil the numbers. decrease, few being met with at a depth of 5 or 6 ft.

The efficiency of many substances, such as farm-yard manure, guanos, bone-meal and all other organic materials, which are spread over or dug or ploughed into the land for the benefit of farm and garden crops, is bound up with the action of these minute living beings. Without their aid most manures would be useless for plant growth. Farm yard manure, guanos and other fertilizers undergo decomposition in soil and become broken down into compounds of simple chemical composition better suited for absorption by the roots of crops, the changes involved being directly due to the activity of bacteria and fungi. Much of the work carried on by these organisms is not clearly understood; there are, however, certain processes which have been extensively investigated and to these it is necessary to refer.

It has been found by experiment that the nitrogen needed by practically all farm crops except leguminous ones is best supplied in the form of a nitrate; the rapid effect of nitrate of soda when used as a top dressing to wheat or other plants is well known to farmers. It has long been known that when organic materials such as the dung and urine of animals, or even the bodies of animals and plants, are applied to soil, the nitrogen within them becomes oxidised, and ultimately appears in the form of nitrate of lime, potash or some other base. The nitrogen in decaying roots, in the dead stem and leaves of plants, and in humus generally is sooner or later changed into a nitrate, the change being effected by bacteria.

That the action of living organisms is the cause of the production of nitrates is supported by the fact that the change does not occur when soil is heated nor when it is treated with disinfectants which destroy or check the growth and life of bacteria. The process resulting in the formation of nitrates in the soil is spoken of as *nitrification*. The steps in the breaking down of the highly complex nitrogenous proteid compounds contained in the humus of soil, or applied to the latter by the farmer in the form of dung and organic refuse generally, are many and varied; most frequently the insoluble proteids are changed by various kinds of putrefactive bacteria into soluble proteids (peptones), these into simpler amido-bodies, and these again sooner or later into compounds of ammonia. The urea in urine is also rapidly converted by the uro-bacteria into ammonium carbonate.

The compounds of ammonia thus formed from the complex substances by many varied kinds of microorganisms are ultimately oxidised into nitrates. The change takes place in two stages and is effected by two special groups of nitrifying bacteria, which are present in all soils.

In the first stage the ammonium compounds are oxidised to nitrites by the agency of very minute motile bacteria belonging to the genus *Nitrosomonas.* The further oxidation of the nitrite to a nitrate is effected by bacteria belonging to the genus *Nitrobacter*. Several conditions must be fulfilled before nitrification can occur. In the first place an adequate temperature is essential; at 5°C or 6°C (40-43°F) the process is stopped, so that it does not go on in winter. In summer, when the temperature is about 24°C(75°F), nitrification proceeds at a rapid rate. The organisms do not carry on their work in soils deficient in air; hence the process is checked in water logged soils. The presence of a base such as lime or magnesia (or their carbonates) is also essential, as well as an adequate degree of moisture in dry soils nitrification ceases.

It is the business of the farmer and gardener to promote the activity of these organisms by good tillage, careful drainage and occasional application of lime to soils which are deficient in this substance. It is only when these conditions are attended to that decay and nitrification of dung, guano, fish-meal, sulphate of ammonia and other manures take place, and the constituents which they contain become available to the crops for whose benefit they have been applied to the land.

Nitrates are very soluble in water and are therefore liable to be washed out of soil by heavy rain. They are, however, very readily absorbed by growing plants, so that in summer, when nitrification is most active, the nitrates produced are usually made use of by crops before loss by drainage takes place. In winter, however, and in fallows loss takes place in the subsoil water.

There is also another possible source of loss of nitrates through the activity of denitrifying bacteria. These organisms reduce nitrates to nitrites and finally to ammonia and gaseous free nitrogen which escapes into the atmosphere. Many bacteria are known which are capable of denitrification, some of them being abundant in fresh dung and upon old straw. They can, however, only carry on their work extensively under anaerobic conditions, as in waterlogged soils or in those which are badly tilled, so that there is but little loss of nitrates through their agency.

An important group of soil organisms are now known which have the power of using the free nitrogen of the atmosphere for the formation of the complex nitrogenous compounds of which their bodies are largely composed. By their continued action soil becomes enriched with nitrogenous material which eventually through the nitrification process becomes available to ordinary green crops. This power of "fixing nitrogen", as it is termed, is apparently not possessed by higher green plants. The bacterium, *Clostridium pasteurianum*, common in most soils, is able to utilise free nitrogen under anaerobic conditions, and an organism known as *Azotobacter chroococcum* and some others closely allied to it, have similar powers which they can exercise under aerobic conditions. For the carrying on of their functions they all need to be supplied with carbohydrates or other carbon compounds which they obtain ordinarily from humus and plant residues in the soil, or possibly in some instances from carbohydrates manufactured by minute green algae with which they live in close union. Certain bacteria of the nitrogenfixing class enter into association with the roots of green plants, the best-known examples being those which are met with in the nodules upon the roots of clover, peas, beans, sainfoin and other plants belonging to the leguminous order.

That the fertility of land used for the growth of wheat is improved by growing upon it a crop of beans or clover has been long recognised by farmers. The knowledge of the cause, however, is due to modern investigations. When wheat, barley, turnips and similar plants are grown, soil upon which they are cultivated becomes depleted of its nitrogen; yet after a crop of clover or other leguminous plants the soil is found to be richer in nitrogen than it was before the crop was grown. This is due to the nitrogenous root residues left in the land.

Upon the roots of leguminous plants characteristic swollen nodules or tubercles are present. These are found to contain large numbers of a bacterium termed *Bacillus radicicola* or *Pseudomonas radicicola*. The bacteria, which are present in almost all soils, enter the root-hairs of their host plants and ultimately stimulate the production of an excrescent nodule, in which they live. For a time after entry they multiply, obtaining the nitrogen necessary for their nutrition and growth from the free nitrogen of the air, the carbohydrate required being supplied by the pea or clover plant in whose tissues they make a home.

The nodules increase in size, and analysis shows that they are exceedingly rich in nitrogen up to the time of flowering of the host plant. During this period the bacteria multiply and most of them assume a peculiar thickened or branched form, in which state they are spoken of as bacteroids. Later the nitrogen-content of the nodule decreases, most of the organisms, which are largely composed of proteid material, becoming digested and transformed into soluble nitrogenous compounds which are conducted to the developing roots and seeds. After the decay of the roots some of the unchanged bacteria are left in the soil, where they remain ready to infect a new leguminous crop.

The nitrogen-fixing nodule bacteria can be cultivated on artificial media, and many attempts have been made to utilise them for practical purposes. Pure cultures may be made and after dilution in water or other liquid can be mixed with soil to be ultimately spread over the land which is to be infected. The method of using them most frequently adopted consists in applying them to the seeds of leguminous plants before sowing, the seed being dipped for a time in a liquid containing the bacteria. In this manner organisms obtained from red clover can be grown and applied to the seed of red clover; and similar inoculation can be arranged for other species, so that an application of the bacteria most suited to the particular crop to be cultivated can be assured.

In many cases it has been found that inoculation, whether of the soil or of the seed, has not made any appreciable difference to the growth of the crop, a result no doubt due to the fact that the soil had already contained within it an abundant supply of suitable organisms. But in other instances greatly increased yields have been obtained where inoculation has been practised.

More or less pure cultures of the nitrogen-fixing bacteria belonging to the *Azotobacter* group have been tried and recommended for application to poor land in order to provide a cheap supply of nitrogen. The application of pure cultures of bacteria for improving the fertility of the land is still in an experimental stage. There is little doubt, however, that in the near future means will be devised to obtain the most efficient work from these minute organisms, either by special artificial cultivation and subsequent application to soil, or by improved methods of encouraging their healthy growth and activity in the land where they already exist.

IMPROVEMENT OF SOILS

The fertility of a soil is dependent upon a number of factors, some of which, such as the addition of fertilizers or manures, increase the stock of available food materials in soil, while others, such as application of clay or humus, chiefly influence the fertility of the land by improving its physical texture.

The chief processes for the improvement of soils are: liming, claying and marling, warping, paring and burning, and green manuring. Most of these more or less directly improve the land by adding to it certain plant food constituents which are lacking, but the effect of each process is in reality very complex. In the majority of cases the good results obtained are more particularly due to the setting free of "dormant" or "latent" food constituents and to the amelioration of the texture of the soil, so that its aeration, drainage, temperature and water-holding capacity are altered for the better.

The material which chemists call calcium carbonate is met with in a comparatively pure state in chalk. It is present in variable amounts in limestones of all kinds, although its whiteness may there be masked by the presence of iron oxide and other coloured substances. Carbonate of lime is also a constituent to a greater or lesser extent in almost all soils. In certain sandy soils and in a few stiff clays it may amount to less than 4%, while in others in limestone and chalk districts there may be 50 to 80% present. Pure carbonate of lime when heated loses 44% of its weight, the decrease being due to the loss of carbon dioxide gas. The resulting white product is termed calcium oxide lime, burnt lime, quicklime, cob lime, or caustic lime. This substance absorbs and combines with water very greedily, at the same time becoming very hot, and falling into a fine dry powder, calcium hydroxide or slaked lime, which when left in the open slowly combines with the carbon dioxide of the air and becomes calcium carbonate, from which we began.

When recommendations are made about liming land it is necessary to indicate more precisely than is usually done which of the three classes of material named above – chalk, quicklime or slaked lime – is intended. Generally speaking the oxide or quicklime has a more rapid and greater effect in modifying the soil than slaked lime, and this again greater than the carbonate or chalk.

Lime in whatever form it is applied has a many-sided influence in the fertility of the land. It tends to improve the tilth and the capillarity of the soil by binding sands together somewhat and by opening up clays. If applied in too great an amount to light soils and peat land it may do much damage by rendering them too loose and open. The addition of small quantities of lime, especially in a caustic form, to stiff greasy clays makes them much more porous and pliable. A lump of clay, which if dried would become hard and intractable, crumbles into pieces when dried after adding to it 2% of lime. The lime causes the minute separate particles of clay to flocculate or group themselves together into larger compound grains between which air and water can percolate more freely.

It is this power of creating a more crumbly tilth on stiff clays that makes lime so valuable to the farmer. Lime also assists in the decomposition of the organic matter or humus in soil and promotes nitrification; hence it is of great value after green manuring or where the land contains much humus from the addition of bulky manures such as farmyard dung. This tendency to destroy organic matter makes the repeated application of lime a pernicious practice, especially on land which contains little humus to begin with. The more or less dormant nitrogen and other constituents of the humus are made immediately available to the succeeding crop, but the capital of the soil is rapidly reduced, and unless the loss is replaced by the addition of more manures the land may become sterile.

Although good crops may follow the application of lime, the latter is not a direct fertilizer or manure and is no substitute for such. Its best use is obtained on land in good condition, but not where soil is poor. When used on light dry land it tends to make the land drier, since it destroys the humus which so largely assists in keeping water in the soil. Lime is a base and neutralises the acid materials present in badly drained meadows and boggy pastures. Weeds, therefore, which need sour conditions for development are checked by liming and the better grasses and clovers are encouraged. It also sets free potash and possibly other useful plant food-constituents of soil. Liming tends to produce earlier crops and destroys the fungus which causes finger-and-toe or club-root among turnips and cabbages.

The particular form in which lime should be applied for the best results depends upon the nature of soil. In practice the proximity to chalk pits or lime kilns, the cost of the lime and cartage, will determine which is most economical. Generally speaking, light poor lands deficient in organic matter will need the less caustic form or chalk, while quicklime will be most satisfactory on the stiff clays and richer soils. On the stiff soils overlying the chalk it was formerly the custom to dig pits through the soil to the rock below. Shafts 20 or 30 ft. deep were then sunk, and the chalk taken from horizontal tunnels was brought to the surface and spread on the land at the rate of about 60 loads per acre. Chalk should be applied in autumn, so that it may be split by the action of frost during the winter. Quicklime is best applied, perhaps, in spring at the rate of one ton per acre every six or eight years, or in larger doses – 4 to 8 tons – every 15 to 20 years. Small dressings applied at short intervals give the most satisfactory results. The quicklime should be placed in small heaps and covered with soil if possible until it is slacked and the lumps have fallen into powder, after which it may be spread and harrowed in. Experiments have shown that excellent effects can be obtained by applying 5 or 6 cwt. of ground quicklime.

Gas-lime is a product obtained from gasworks where quicklime is used to purify the gas from sulphur compounds and other objectionable materials. It contains a certain amount of unaltered caustic lime and slacked lime, along with sulphates and sulphides of lime, some of which have an evil odour. As some of these sulphur compounds have a poisonous effect on plants, gas-lime cannot be applied to land directly without great risk or rendering it incapable of growing crops of any sort – even weeds

– for some time. It should therefore be kept a year or more in heaps in some waste corner and turned over once or twice so that the air can gain access to it and oxidise the poisonous ingredients in it.

Many soils of a light sandy or gravelly or peaty nature and liable to drought and looseness of texture can be improved by the addition of large amounts of clay of an ordinary character.

Similarly soils can be improved by applying to them marl, a substance consisting of a mixture of clay with variable proportions of lime. Some of the chalk marls, which are usually of a yellowish or dirty grey colour, contain clay and 50 to 80% of carbonate of lime with a certain proportion of phosphate of lime. Such a material would not only have an influence on the texture of the land but the lime would reduce the sourness of the land and the phosphate of lime supply one of the most valuable of plant food constituents. The beneficial effects of marls may also be partially due to the presence in them of available potash.

Typical clay-marls are tenacious, soapy clays of yellowish-red or brownish colour and generally contain less than 50% of lime. When dry they crumble into small pieces which can be readily mixed with the soil by ploughing. Many other kinds of marls are described; some are of a sandy nature, others stony or full of the remains of small shells. The amount and nature of the clay or marl to be added to the soil will depend largely upon the original composition of the latter, the lighter sands and gravel requiring more clay than those of firmer texture. Even stiff soils deficient in lime are greatly improved in fertility by the addition of marls. In some cases as little as 40 loads per acre have been used with benefit, in others 180 loads have not been too much. The material is dug from neighbouring pits or sometimes from the fields which are to be improved, and applied in autumn and winter. When dry and in a crumbly state it is harrowed and spread and finally ploughed in and mixed with the soil.

On some of the strongest land it was formerly the practice to add to and plough into it burnt clay, with the object of making the land work more easily. The burnt clay moreover carried with it potash and other materials in a state readily available to the crops. The clay is dug from the land or from ditches or pits and placed in heaps of 60 to ioo loads each, with faggot wood, refuse coals or other fuel. Great care is necessary to prevent the heaps from becoming too hot, in which case the clay becomes baked into hard lumps of brick-like material which cannot be broken up. With careful management, however, the clay dries and bakes, becoming slowly converted into lumps which readily crumble into a fine powder, in which state it is spread over and worked into the land at the rate of 40 loads per acre.

The paring and burning of land, although formerly practised as an ordinary means of improving the texture and fertility of arable fields, can now only be looked upon as a practice p to be adopted for the purpose of bringing rapidly into cultivation very foul leys or, land covered with a coarse turf. The practice is confined to poorer types of land, such as heaths covered with furze and bracken or fens and clay areas smothered with rank grasses and sedges. To reduce such land to a fit state for the growth of arable crops is very difficult and slow without resort to paring and burning. The operation consists of paring off the tough sward to a depth of 1 to 2 inch just sufficient to effectually damage the roots of the plants forming the sward and then, after drying the sods and burning them, spreading the charred material and ashes over the land. The turf is taken off either with the breast plough – a paring tool pushed forward from the breast or thighs by the workman – or with specially constructed paring ploughs or shims.

The depth of the sod removed should not be too thick or burning is difficult and too much humus is destroyed unnecessarily, nor should it be too thin or the roots of the herbage are not effectually destroyed. The operation is best carried out in spring and summer. After being pared off the turf is allowed to dry for a fortnight or so and is then placed in small heaps a yard or two wide at the base, a little straw or wood being put in the middle of each heap, which is then lighted. As burning proceeds more turf is added to the outside of the heaps in such a manner as to allow little access of air. Every care should be taken to burn and char the sod thoroughly without permitting the heap to blaze. The ashes should be spread as soon as possible and covered by a shallow ploughing. The land is then usually sown with some rapidly growing green crop, such as rape, or with turnips.

Paring and burning improves the texture of clay lands, particularly if draining is carried out at the same time. It tends to destroy insects and weeds, and gets rid of acidity of soil. No operation brings old turf into cultivation so rapidly. Moreover, the beneficial effects are seen in the first crop and last for many years. Many of the mineral plant food-constituents locked up in the coarse herbage and in the upper layers of soil are made immediately available to crops. The chief disadvantage is the loss of nitrogen which it entails, this element being given off into the air in a free gaseous state. It is best adapted for application to clays and fen lands and should not be practised on shallow light sands or gravelly soils, since the humus so necessary for the fertility of such areas is reduced too much and the soil rendered too porous and liable to suffer from drought.

Many thousands of acres of low-lying peaty and sandy land adjoining the tidal rivers which flow into the Humber have been improved by a process termed "warping". The warp consists of fine muddy sediment which is suspended in the tidal river water and appears to be derived from material scoured from the bed of the Humber by the action of the tide and a certain amount of sediment brought down by the tributary streams which join the Humber some distance from its mouth. The field or area to be warped must lie below the level of the water in the river at high tide. It is first surrounded by an embankment, after which the water from the river is allowed to flow through a properly constructed sluice in its bank, along a drain or ditch to the land which is prepared for warping. By a system of carefully laid channels the water flows gently over the land, and deposits its warp with an even level surface. At the ebb of the tide the more or less clear water flows back again from the land into the main river with sufficient force to clean out any deposit which may have accumulated in the drain leading to the warped area, thus allowing free access of more warpladen water at the next tide. In this manner poor peats and sands may be covered with a large layer of rock soil capable of growing excellent crops.

The amount of deposit laid over the land reaches a thickness of two or three feet in one season of warping, which is usually practised between March and October, advantage being taken of the spring tides during these months. The new warp is allowed to lie fallow during the winter after being laid out in fouryard "lands" and becomes dry enough to be sown with oats and grass and clover seeds in the following spring. The clover-grass ley is then grazed for a year or two with sheep, after which wheat and potatoes are the chief crops grown on the land.

Green manures are crops which are grown especially for the purpose of ploughing into the land in a green or actively growing state. The crop during its growth obtains a considerable amount of *Green* carbon from the carbon dioxide of the air, and builds it up *Manuring* into compounds which when ploughed into the land become humus. The carbon compounds of the latter are of no direct nutritive value to the succeeding crop, but the decaying vegetable tissues very greatly assist in retaining moisture in light sandy soils, and in clay soils also have a beneficial effect in rendering them more open and allowing of better drainage of superfluous water and good circulation of fresh air within them. The ploughing-in of green crops is in many respects like the addition of farm-yard manure.

Their growth makes no new addition of mineral food-constituents to the land, but they bring useful substances from the subsoil nearer to the surface, and after the decay of the buried vegetation these become available to succeeding crops of wheat or other plants. Moreover, where deep-rooting plants are grown the subsoil is aerated and rendered more open and suitable for the development of future crops.

The plants most frequently used are white mustard, rape, buckwheat, spurry, rye, and several kinds of leguminous plants, especially vetches, lupins and serradella. By far the most satisfactory crops as green manures are those of the leguminous class, since they add to the land considerable amounts of the valuable fertilising constituent, nitrogen, which is obtained from the atmosphere. By nitrification this substance rapidly becomes available to succeeding crops. On the light, poor sands of Saxony Herr Schultz, of Lupitz, made use of serradella, yellow lupins and vetches as green manures for enriching the land in humus and nitrogen, and found the addition of potash salts and phosphates very profitable for the subsequent growth of potatoes and wheat.

He estimated that by using leguminous crops in this manner for the purpose of obtaining cheap nitrogen he reduced the cost of production of wheat more than 50%. The growing crops should be ploughed in before flowering occurs; they should not be buried deeply, since decay and nitrification take place most rapidly and satisfactorily when there is free access of air to the decaying material. When the crop is luxuriant it is necessary to put a roller over it first, to facilitate proper burial by the plough. The best time for the operation appears to be late summer and autumn.

The influence of different kinds of soil as a factor in the production of disease requires to be considered, in regard not only to the nature and number of the microorganisms they contain, but also to the amount of moisture and air in them and their capacity for heat. The moisture in soil is derived from two sources—the rain and the ground-water. Above the level of the ground-water the soil is kept moist by capillary attraction and by evaporation of the water below, by rainfall, and by movements of the ground-water; on the other hand, the upper layers are constantly losing moisture by evaporation from the surface and through vegetation.

When the ground water rises it forces air out of the soil; when it falls again it leaves the soil moist and full of air. The nature of the soil will largely influence the amount of moisture which it will take up or retain. In regard to water, all soils have two actions, namely, permeability and absorbability. Permeability is practically identical with the speed at which percolation takes place; through clay it is slow, but increases in rapidity through marls, loams, limestones, chalks, coarse gravels and fine sands, reaching a maximum in soil saturated with moisture.

The amount of moisture retained depends mainly upon the absorbability of soil, and as it depends largely on capillary action it varies with the coarseness or fineness of the pores of the soil, being greater for soils which consist of fine particles. The results of many analyses show that the capacity of soils for moisture increases with the amount of organic substances present; decomposition appears to be most active when the moisture is about 4%, but can continue when it is as low as 2%, while it appears to be retarded by any excess over 4%.

Above the level of the ground-water all soils contain air, varying in amount with the degree of looseness of the soil. Some sands contain as much as 50% of air of nearly the same composition as atmospheric air. The oxygen, however, decreases with the depth, while the carbon dioxide increases. Among the most noteworthy workers at the problems involved in the question of the influence of soil in the production of disease we find yen Foder, Pettenkofer, Levy, Fleck, von Naegeli, Schleesing, Muntz and Warrington.

The study of epidemic and endemic diseases generally has brought to light an array of facts which very strongly suggest that an intimate association exists between soil and the appearance and propagation of certain diseases; but although experiments and observations allow this view to be looked upon as well established, still the precise role played by soil in an aetiological respect is by no means so well understood as to make it possible to separate the factors and dogmatise on their effects.

The earliest writers upon cholera emphasised its remarkable preference for particular places; and the history of each successive epidemic implies, besides an importation of the contagion, certain local conditions which may be either general sanitary defects or peculiarities of climate and soil.

The general evidence indicates that the specific bacteria of cholera discharges are capable of a much longer existence in the superficial soil layers than was formerly supposed; consequently it is specially necessary to guard against pollution of soil, and through it against the probable contamination of both water and air.

The evidence, however, is not sufficiently strong to warrant a universal conclusion, the diffusion of cholera appearing to be largely dependent upon other factors than soil states. Again, all accounts of diphtheria show a tendency on the part of the disease to recur in the same districts year after year.

The questions naturally suggest themselves – Are the reappearances due to a revival of the contagion derived from previous outbreaks in the same place, or to some favouring condition which the place offers for the development of infection derived from some other quarter; and have favouring conditions any dependence upon the character and state of the soil?

Greenhow in 1858 stated that diphtheria was especially prevalent on cold, wet soils, and Airy in 1881 described the localities affected as "for the most part cold, wet, clay lands."

An analysis of the innumerable outbreaks in various parts of Europe indicates that the geological features of the affected districts play a less important part in the incidence of the disease than soil dampness. In this connexion it is interesting to note the behaviour of the diphtheritic contagion in soil. Experiments show that pure cultures, when mixed with garden soil constantly moistened short of saturation and kept in the dark at a temperature of 14°C, will retain their vitality for more than ten months; from moist soil kept at 26°C they die out in about two months; from moist soil at 30°C. in seventeen days; and in dry soil at the same temperature within a week.

In the laboratory absolute soil dryness is as distinctly antagonistic to the vitality of the diphtheria bacillus as soil dampness is favourable. Both statistically and experimentally we find that a damp soil favours its life and development, while prolonged submersion and drought kill it. We may consider that, in country districts, constant soil moisture is one of the chief factors; while in the case of urban outbreaks mere soil moisture is subsidiary to other more potent causes.

Chapter 2

Microbiological Studies of Soil

THE SOIL ECOSYSTEM

Understanding the role of soil in the farm ecosystem, and knowing how to manage the land, are critical and difficult tasks facing the organic farmer. Soil's biological and electrochemical processes cannot be observed directly since they take place at a microscopic and sub-molecular level. Changes in fertility, tilth and structure may take years to become evident. Early indicators are subtle, and the farmer must be a keen observer to spot them. Much of the material will be familiar; the key difference is the recognition of the vital role soil microorganisms play in recycling, releasing, and storing plant nutrients. Organic farmers use techniques that support and enhance the biological life of the soil, which in turn nurtures the crop and maintains soil structure.

Soil Biology

Most of the farm's life exists out of sight, beneath the surface of soil. Billions of organisms inhabit the upper layers of soil, where they break down dead organic matter, releasing the nutrients necessary for plant growth. The microorganisms include bacteria, actinomycetes, algae and fungi.

Macroorganisms include earthworms and arthropods such as insects, mites and millipedes. Each group plays a role in the soil ecosystem and can assist the organic farmer in producing a healthy crop. Microorganisms can be grouped according to their function: free-living decomposers convert organic matter into nutrients for plants and other microorganisms, rhizosphere organisms are symbiotically associated with the plant roots and free-living nitrogen fixers.

The Decomposers

In an undisturbed soil, leaves and other organic debris accumulate on the surface, where they are broken down by the decomposers. Aerobic bacteria and certain small animals begin the process. These organisms are joined by actinomycetes and fungi. Mites, springtails, small insects, other arthropods and earthworms assist the process by consuming, mixing and transporting materials. The rate of decomposition is affected by soil temperature, moisture and food availability. The main by-products of the decomposition process are soluble plant nutrients and microbial remains that bind soil particles together, giving a stable crumb structure. Since biological activity is greatest when soil is warm, nutrient availability is highest during summer, when crop needs are greatest. The decomposers are most active in the upper layer of soil, i.e., the top 8 cm. (3 in.). Organic farmers incorporate organic matter into the surface layers when conditions are favourable to stimulate decomposition and thereby provide plant nutrients.

Rhizosphere Organisms

Plant roots leak or exude a large number of organic substances and continually slough off root caps into soil. These materials are food for the many microorganisms living in a zone of intense biological activity near the roots called the rhizosphere. Bacteria benefit most from the food supplied in the rhizosphere and may form a continuous film around the root. Roots form the microbial highways of the soil. Other microorganisms liberate nutrients from the clay and humus colloids (a colloid is a mass of fine particles).

Symbiotic Organisms in the Rhizosphere

The best-known symbiotic relationship occurs between nitrogen-fixing *Rhizobia* bacteria and legumes. The *Rhizobia* inhabit small pea-like lumps on the roots, extracting carbohydrates from the plant and providing the plant with soluble nitrogen compounds synthesized from nitrogen gas in the soil atmosphere. Mycorrhizal fungi have similar symbiotic relationships with the roots of many plants. By extending the surface area of the roots by as much as 400 times, the fungi help the plant with the absorption of water and nutrients and with its ability to withstand heat and drought. These symbiotic relationships begin at germination when the young sprout exudes toxins to kill pathogens and hormones to attract beneficial organisms.

Soil Animals

The most important group of larger soil animals are the earthworms, of which there may be a dozen species in a healthy Canadian soil. Earthworms perform the final task of humification — the conversion of decomposed organic matter to stable humus colloids — and mix the humus with material from the lower soil horizons. The digestive tract of the earthworm has a remarkable capacity to literally alter the chemical and physical nature of soil. Earthworms are major agents in the process of soil creation through the formation of clay-humus complexes and they play a key role in the management of calcium. By inoculating their castings with intestinal flora, earthworms distribute microbial populations throughout the soil. Earthworms can increase the availability of phosphorus from rock phosphate by 15-39 per cent.

They act as mini-subsoilers, their burrows increasing soil aeration, drainage and porosity. In the process of burrowing, earthworms mix the subsoil with the topsoil and deposit their nutrient-rich castings on or near the soil surface. The presence of a large earthworm population indicates good soil fertility. They can be encouraged by adding lime when needed to correct soil acidity and organic matter to provide the worms with food. Note that the red wriggler, or manure worm, prefers an environment higher in organic matter and cannot survive in most soils; inoculating a field with these worms will not improve soil fertility.

Mites are the most abundant of the soil arthropods. Most mites are beneficial, feeding on microorganisms and other small animals. They assist with decomposition by browsing on preferred fungi, thus preventing any one species from becoming dominant, and by transporting the spores through soil. Springtails perform similar functions. Larger arthropods, slugs and snails burrow through soil and feed on dead plant material. By maintaining a suitable environment for the hundreds of species of soil creatures, large and small, organic farmers provide their crops with an abundant supply of plant nutrients.

SOIL CHEMISTRY (FERTILITY)

Organic Matter and Humus

Organic matter (OM) is the term used to describe the component of soil made up of the remains, residues or waste products of any living organism. OM comes primarily from plant residue, but also includes soil microorganisms and animal remains. The amount of OM in a soil depends on its type and how it is managed. OM content can range from one per cent, in the case of a sandy soil in which no special management practices have been used to build OM, to more than 30 per cent in a muck soil.

Soil life depends on the continual replenishment of OM. Most organic farming practices, such as crop rotation, composting, green manuring and keeping the soil covered, help to increase the soil's OM and hence its biological activity. Including a three- to five-year grass-legume sod in the rotation is an effective way of increasing OM because losses are minimised when there is no tillage. It is important to understand that OM alone does not guarantee fertility or biological activity. Peat moss, for example, is made up entirely of OM but contains few nutrients. Excessive moisture will produce anaerobic conditions in which OM will rot and will favour the development of pathogens that may infect the crop. Soil must be managed so that the OM produces the intended results, namely, an increase in available plant nutrients, improved soil structure, and increased nutrient reserves.

Effective Humus

When fresh OM is added to soil, the microorganisms immediately start to decompose it. The partially-decomposed residue is called effective humus. It holds nutrients by absorption, releasing them to plants as needed and prevents their loss by leaching. Microorganisms that decompose residues with a high carbon content will utilise some of the available nitrogen, making it temporarily unavailable to a crop that is seeded immediately afterwards.

Stable Humus

Stable humus is the final product of the decomposition process. It can be recognised by its dark colour, crumbly or slightly gelatinous texture and characteristic "earthy" smell. Stable humus, or colloidal humus, provides long-term nutrient reserves and improves soil structure and cation-exchange capacity.

Benefits of humus:

- Supplies nutrients, especially nitrogen (N), phosphorus (P) and sulphur (S), when the plant needs them;
- Holds nutrients, thereby reducing nutrient leaching;
- Binds soil particles together, stabilising loose soils against erosion;
- Increases the friability of heavy soils; and
- Improves porosity, thereby facilitating air and water movement, and increases soil's water-holding capacity.

CATION-EXCHANGE CAPACITY (CEC)

Plants obtain many of their nutrients from soil by an electrochemical process called cation exchange. This process is the key to understanding soil fertility. Cation exchange requires very small particles with a large surface area to hold electrically-charged ions. Humus colloids are ideal; clay colloids also have a good CEC, but sand particles are too big. The finely-divided platelets of the humus and clay colloids produce a large surface area — one gramme of the clay mineral bentonite has been estimated to have a surface area of 800 square metres. The surfaces are coated with a thin film of water, which contains dissolved nutrients. Each platelet has an extra electron, which gives it a negative charge. This negative charge attracts positively-charged nutrient ions from the nutrient solution such as ammonium (NH_4^+), calcium (Ca^{++}), magnesium (Mg^{++}) and potassium (K^+). These nutrient ions can be absorbed by the plant root, by exchanging them for other ions such as hydrogen (H^+). Many soil micro-organisms carry a negative charge, which enables them to attract nutrients, and to move freely about the humus and clay colloids.

CEC measures the quantity of potentially-available cation nutrients that are in a stable, accessible form. It is measured in milliequivalents (me) per 100 gms of soil. Typical values are 6.3 me/100g for sand and 27.2 me/100g for clay/loam. The higher the CEC, the greater the potential fertility of the soil. This is why clay soils tend to be more fertile than sandy soils, and why the fertility of sandy soils can be improved by the addition of clay and humus. The cation-exchange process can however only store and release positively-charged nutrients; the availability of nutrients in anion form, such as phosphorus and sulfur is not affected by CEC. Soil organisms play a key role in conserving and releasing these nutrients.

Soil pH and the Role of Calcium

The term pH refers to the acidity or alkalinity of a soil. It is important because it influences soil nutrient availability and biological activity. pH ranges from 0-14. A pH level below 7 (the neutral point) is acidic, and above 7 is alkaline. Soil pH ranges from 4-9; fertile soils are usually between 6.0 and 7.0.

Acid soils have, by definition, a large number of free H^+ ions. Acidity reduces bacterial activity and therefore decomposition and nutrient release. Nitrogen-fixing *Rhizobia* and legumes generally do not do well in acid soils. Excess H+ ions displace nutrient cations attached to soil colloids, thus depleting soil's nutrient reserves. An acid soil may, therefore, have a high CEC but be low in fertility.The addition of crushed limestone ($CaCO_3$) corrects an acid soil. An acid soil with a high CEC needs a greater amount of limestone than a low CEC soil of the same pH, because of the very much greater number of reserve H+ ions held in the soil with the high CEC.

Lime not only corrects soii pH, it also supplies the plant nutrient calcium. Its double electrical charge, Ca^{++}, lets it function as a link, binding clay and humus colloids together in clay-humus complexes. The resulting soil has improved structure, is less subject to erosion and has improved nutrient-holding capacities. Dolomitic limestone functions in a similar way to calcitic limestone, but in addition contains magnesium (Mg^{++}). It should only be used in areas that are low in magnesium. If magnesium levels are high compared with calcium it will have adverse effects on the crops and on the breakdown of organic residues in the soil.Excessively alkaline soils have few free H^+ ions and an excess of sodium (Na^+) ions. Biological activity is suppressed and associated nutrient availability decreased. Additional problems include destruction of OM, saline seepage, soil crusting and the accumulation of toxic levels of sodium, selenium and other minerals. Alkalinity can be reduced somewhat by the addition of gypsum (calcium sulphate) or, in extreme circumstances, sulfur. Gypsum is used to reduce magnesium and supply calcium and sulphur without raising the pH.

BIOGEOCHEMICAL CYCLES

The Earth is a closed system for matter, except for small amounts of cosmic debris that enter the Earth's atmosphere. This means that all the elements needed for the structure and chemical processes of life come from the elements that were present in the Earth's crust when it was formed billions of years ago. This matter, the building blocks of life, continually cycle through the Earth's systems, the atmosphere, hydrosphere, biosphere, and lithosphere, on time scales that range from a few days to millions of years. These cycles are called biogeochemical cycles, because they include a variety of biological, geological, and chemical processes.

Many elements cycle through ecosystems, organisms, air, water, and soil. Many of these are trace elements. Other elements, including carbon, nitrogen, oxygen, hydrogen, sulfur, and phosphorus are critical components of all biological life. Together, oxygen and carbon account for 80 per cent of the weight of human beings. Because these elements are key components of life, they must be available for biological processes. Carbon, however, is relatively rare in the Earth's crust, and nitrogen, though abundant in the atmosphere, is in a form that is not useable by living organisms. The biogeochemical cycles transport and store these important elements so that they can be used by living organisms.

Each cycle takes many different pathways and has various reservoirs, or storage places, where elements may reside for short or long periods of time. Each of the chemical, biological, and geological processes varies in their rates of cycling. Some molecules may cycle very quickly depending on the pathway. Carbon atoms in deep ocean sediments may take hundreds to millions of years to cycle completely through the system. An average water molecule resides in the atmosphere for about ten days, although it may be transported many miles before it falls back to the Earth as rain.

How fast substances cycle depends on its chemical reactivity and whether or not it can be found in a gaseous state. A gaseous phase allows molecules to be transported quickly. Phosphorous has no gaseous phase and is relatively uncreative, so it moves very slowly through its cycle. Phosphorus is stored in large amounts in sediment in the oceans or in the

Earth's crust and is recycled back to the surface only over very long periods of time through upwelling of ocean waters or weathering of rocks.

Biogeochemical cycles are subject to disturbance by human activities. Humans accelerate natural biogeochemical cycles when elements are extracted from their reservoirs, or sources, and deposited back into the environment (sinks). For example, humans have significantly altered the carbon cycle by extracting and combusting billions of tons of hydrocarbons in fossil that were buried deep in the Earth's crust, in addition to clearing vegetation that stores carbon.

Global release of carbon through human activities has increased from 1 billion tons per year in 1940 to 6.5 billion tons per year in 2000. About half of this extra carbon is taken up by plants and the oceans, while the other half remains in the atmosphere. In addition to carbon cycle, humans have altered the nitrogen and phosphorus cycles by adding these elements to croplands as fertilizers, which has contributed to over-fertilization of aquatic ecosystems when excess amounts are carried by runoff into local waterways.

Nitrogen Cycle

The vegetative growth of plants (leaves, stems, and roots) is especially dependent on nitrogen. The atmosphere contains 78 per cent nitrogen by volume, yet it is the element that most often limits plant growth. Plants cannot use gaseous nitrogen, but require nitrogen in the form of nitrate (NO_3^-) or ammonium (NH_4^+). Atmospheric nitrogen is converted into NO_3^- and NH_4^+ in the soil by nitrogen fixation, which is performed by certain soil microorganisms. These include the symbiotic *Rhizobia* bacteria associated with legumes, and the non-symbiotic bacteria *Clostridium* and *Azoterbacter* which are free-living in the soil. Once gaseous nitrogen is incorporated into plant material as proteins and amino acids, it may be recycled many times through the activity of the soil decomposers.

Young plants are especially rich in nitrogen and, when they are incorporated in the surface layers of soil as green manure, this nitrogen is released by biological activity. The ammonium (NH_4^+) ions can be stored on the clay-humus complex for long periods. The nitrate ions (NO_3^-) are subject to leaching if not taken up by the crop. Deficiencies of nitrogen may occur not because there is not enough entering the system but because of the way it cycles round the system. Cycling is increased by maximising biological activity which is determined by the way different components of the system, such as residues, manure, weeds and drainage, are managed.

Carbon Cycle

Carbon is the building block of life. Plants obtain carbon from atmospheric carbon dioxide (CO_2) through photosynthesis, during which the chloroplasts in the plant cells convert CO_2 to carbohydrates. It is the cycling of carbon from the atmosphere through plants and algae, to animals and microorganisms and back to the atmosphere, that maintains earth's atmosphere and climate in its current balance. The greenhouse effect, or warming of the planet, is a consequence of an excess of atmospheric CO_2 caused by deforestation (reduced CO_2 consumption) and compounded by excessive fossil fuel energy use (increased CO_2 production). Keeping the soil covered with growing plants can make a contribution to reducing global warming. Carbon is a critical element in the formation of stable humus. The carbon : nitrogen (C:N) ratio of the organic matter supplied to the soil is a controlling factor in this process. A ratio of about 20:1 is considered ideal.

If greater amounts of carbon are present, decomposition slows as microorganisms become nitrogen-starved and compete with the plants for available nitrogen. Nitrate nitrogen practically disappears from soil because microbes need nitrogen to build their tissues. If there is too much soil nitrogen, the decomposers produce soluble nutrients in the form of effective humus, but little stable humus. These conditions can give the advantage to weeds rather than the crop. A good C: N ratio will result in the formation of both effective humus and stable humus. As decay occurs, the C: N ratio of the plant material decreases since carbon is being lost as CO_2, and nitrogen is conserved. This process continues until the micro-organisms run out of easily-oxidised carbon. The exuded, undecomposed carbon persists as stable humus.

Phosphorus Cycle

Phosphorus (P) is important in plant-cell division and growth. It is a difficult nutrient to manage because, although abundant in soil, it is often in a form unavailable to plants. In acidic soils (pH below 5) the phosphorus gets tied up with iron and aluminum, and in alkaline soils (pH above 7) it gets tied up with calcium. Even with a favourable pH, phosphorus readily becomes immobilised by other soil minerals. Phosphorus anions may also be physically trapped in the clay-humus complex. Phosphorus is lost from soils through soil erosion, often at a greater rate than it can be replaced from the underlying subsoils.

It accumulates in lakes and slow-flowing rivers, causing eutrophication. The elimination of soil erosion is the first step in phosphorus conservation. The addition of powdered rock phosphate or colloidal phosphate is a precautionary measure which, used in conjunction with the biological measures described below, can avoid phosphorus deficiency. The release of P to plants depends on soil biological activity, particularly that of certain bacteria and mycorrhizal fungi. Soil acids, produced by these microorganisms and by OM decomposition, release phosphates. Phosphorus availability is therefore dependent on the maintenance of high levels of biological activity and stable humus in soil. Under these conditions, phosphorus is continually recycled through the processes of OM decay. Some plants produce acidity around their roots which assists in the uptake of P; examples of these are legumes actively fixing nitrogen, rapeseed, oilradish and buckwheat.

Potassium Cycle

Potassium (K) is important as an enzyme activator in plants. It is involved in facilitating membrane permeability and translocation of sugars. Potassium is also needed for photosynthesis, fruit formation, winter hardiness, disease resistance, and amino acid and protein formation. Potassium builds plant stalk strength. It does not, however, form a permanent part of plant tissues, but is translocated to the stems and roots during ripening.

Thus, potassium is readily available in crop residues — roots, straw and corn stalks. Very little potassium is removed with a grain crop at harvest if the straw is left on the field. Repeated cutting for hay or silage without returning potash in the form of manure or crop residues will quickly induce K deficiency. Soil potassium is present in minerals that dissolve slowly, thereby limiting its availability. Potassium availability is regulated by cation exchange.

Potassium leaching increases as the amounts of clay and humus decrease and therefore may be a problem in sandy soils. A deep-rooting green manure will help prevent losses. Increased biological activity and colloidal humus formation will increase potassium availability by enhancing the CEC in soil. The addition of powdered basalt, green sand and clay minerals has been found to correct potassium deficiencies in a biologically active soil. Manure is a good source of K if care has been taken to minimise leaching during storage. It has been reported that in some organic systems, low available potash levels, according to soil analyses, are not necessarily associated with plant deficiencies or lower yields. This may be because available K is immediately taken up by the growing plant.

Micronutrients

About one hundred elements have been found in living plants. Carbon, hydrogen, and oxygen are the most abundant and are derived from water, oxygen and carbon dioxide. The nutrients N, P, K, calcium and magnesium. Of the other elements, we know that sulfur, iron, copper, manganese, zinc, molybdenum, boron and chlorine are required by plants in trace amounts. They are not constituents of the plant structure, but contribute to plant growth and development.

Other elements, such as iodine, are essential to the animals that eat the plants. Deficiencies occur in soils that lack an inherent source of an element, or they can be caused by an imbalance in soil pH. Conversely, if certain micronutrients exceed trace levels, they can be toxic to plants. The range between deficiency and excess is very small. Therefore, micronutrients should not be applied unless a deficiency is shown by leaf analysis or by visible plant symptoms.

Micronutrients are best applied via compost, or by a foliar spray. Either of these methods is preferable to applying a trace mineral directly to soil. In a biologically-active soil with good CEC and balanced pH, micronutrient deficiencies are rare. Products based on seaweed contain more than 80 elements, and organic farmers feed kelp meal as mineral supplement to their livestock, or incorporate small amounts of kelp products into compost as a precautionary measure against micronutrient deficiency.

WATER, AIR AND DRAINAGE

Fundamental to soil ecology is the cycling of water to soil through precipitation and its return to the air through evaporation and transpiration. Biological activity is dependent upon the balance of air and water in the soil. Too much water causes aerobic decomposition to cease and anaerobic bacteria to take over, with damaging effects. For example, nitrification, or the breakdown of nitrate nitrogen to gaseous nitrogen, occurs as a result of anaerobic biological activity in soil. Too little water also causes biological activity to slow down and hence reduces the availability of nutrients.The water available to plants is the moisture held mostly by capillarity in small soil pores.

A soil with a large number of small pores, such as a clay-loam, will withstand drought much better than a sandy soil, which has few capillary pores. Large pores allow drainage and air flow that supplies oxygen and nitrogen for root and microbial growth. Both types of pore space are important for soil fertility, and both can be maintained and enhanced by the addition of organic matter and humus to soil.

An ideal soil has a high infiltration rate, and fairly slow hydraulic conductivity. The infiltration rate is the rate at which water soaks into the ground; if the infiltration rate is slower than the rate of precipitation, the excess water will become surface run-off, with attendant erosion and pollution hazards. Hydraulic conductivity is the rate at which water drains through a saturated soil. This action transports nutrients from the surface layers to the rhizosphere. If the hydraulic conductivity is too fast, nutrients will be leached out of the soil and groundwater may become polluted.

Organic matter in the form of cover crops or mulch improves the infiltration rate. When converted into humus through biological activity, organic matter can lower the hydraulic conductivity of sandy soils.Wet soils, if caused by high groundwater levels, tend to be unsuitable for organic field crops and are often better left as permanent pasture, or allowed to revert to natural habitat. If the water problem is caused by compaction or hardpan, chisel plowing or subsoiling may correct the situation. Earthworms, and crops with long tap roots such as alfalfa, can then help to maintain the field in improved condition. Solutions such as ditching or tile drainage should be very carefully assessed for their environmental implications.

PHYSICAL PROPERTIES OF SOIL

Soil Structure

The term soil structure is used to describe the way soil particles are grouped into aggregates. Soil structure is affected by biological activity, organic matter, cultivation and tillage practices. Soil fertility and structure are closely related. In an organic production system soil management techniques are designed to enhance soil structure. An ideal soil structure is often described as granular or crumb-like. It provides for good movement of air and water through a variety of different pore sizes.

Plant roots extend down and soil animals, including small earthworms, travel through the spaces between the aggregates. An ideal soil structure is also stable and resistant to erosion. The clay-humus complex, in combination with adequate calcium which helps to bind the aggregates together, forms the basis of this structure. The glutinous by-products of soil bacteria and the hair-like threads of actinomycete and fungi mycelium add to soil stability. Plant roots also play a role in maintaining soil structure. All tillage operations change soil structure.

Excessive cultivation, especially for seedbed preparation, can harm soil structure. Working clay soils when wet leads to compaction and subsequent soil puddling. Soil is easily puddled by rain, easily eroded and will have poor aeration. Tillage, when too dry, shatters the aggregates. Careful cultivation, growing sod crops and returning crop residues can enhance soil structure. Organic matter and the humification process improve structural stability, and can rebuild degraded soil structures. Therefore it is vital to return organic material to the soil and to maintain its biological activity.

Tilth and Tillage

Tilth is the term used by farmers to describe how easy it is to till the soil. It is determined by soil structure, presence or absence of hard-pans, soil moisture and aeration. Tilth determines soil's fitness as a seedbed, especially for root penetration and shoot emergence. However, if the deeper soil layers are compacted or cemented, plant roots will be prevented from getting to the stored water in these layers and plant growth will be affected regardless of upper soil tilth. Tillage should be carried out under conditions that preserve good tilth, that is, when soil moisture conditions are optimum and there is enough water to allow separation of the soil aggregates, but not so much as to induce puddling or compaction. Soil should not stick to your boots when you walk on it and it should break easily and crumble at the deepest depth it is being tilled. This rule is more crucial for fine-textured (clay) soils than for coarse-textured (sandy) soils. A tillage system should work residues into the top 8 cm of the soil where it can be digested by the micro-organisms.

It should also leave some residue on the surface to reduce erosion potential. Annual use of the moldboard plow can create a hardpan and bury organic matter and living topsoil in an anaerobic zone. On many farms its use has been replaced by the chisel plow which loosens, aerates and mixes soil without burying all of the crop residue. If wisely used on soil in good tilth, the moldboard plow need not create problems and it is still useful to turn a heavy sod. However, plowing should be kept as shallow as possible. Overuse of offset discs in the spring can lead to compaction problems and, in some cases, the S-tine cultivator is more appropriate for seed bed preparation. In recent years, equipment modifications and new combination tools have been developed to minimise the adverse effects of tillage on soil structure and to reduce the number of tillage operations required.

Soil Texture

Soil texture is a classification system based on mineral particle size. It is a relatively permanent feature of soil that does not change appreciably over a human lifetime. Soils are classified according to the percentages of oven-dry weights of sand, silt and clay. For example, a sandy soil is composed principally of large sand particles, whereas a loam contains more or less equal amounts of clay, sand and silt. Organic matter is excluded from the texture classification. Soils with a high silt content and those with a high clay content have greater capacities for retaining water and available nutrients than sandy soils.

By adding small amounts of clay minerals to the soil and by encouraging the activities of earthworms to reduce the size of soil mineral particles, organic farmers can modify soil texture to a small degree, but the greatest effect of these amendments is on structure. Soil evaluation is an ongoing process for the organic farmer. Regular observation of the crops and of weed growth provides vital information. The simple act of digging a hole in the field can reveal the following information, which should be recorded:

- Soil profile, which describes the depth and colour of the different soil horizons, or layers;

- Soil structure, including stoniness and hardpan formation at the various soil horizons;
- Earthworm populations and other soil life; and
- Root structures, noting whether roots enter the soil structure, follow fissures made by a chisel plow or subsoiler, or are obstructed in any way.

The information gained from test holes can be used to find out why one part of a field yields differently from another and to compare soil conditions from one year to the next. The plant populations under these soil conditions should be described at the same time, including the density, vigour and composition of the weed population.

Soil Tests

Conventional soil tests are useful indicators. Soil samples should be taken at the same time each year, preferably under the same conditions. The results, compared from year to year, enable the farmer to evaluate the effectiveness of the management practices used and determine what changes are required. It is also important to use the same testing laboratory because different procedures can give different results. Most labs give information on texture, pH, phosphorus, potassium and magnesium but other information useful to organic farmers such as OM, CEC, calcium and micronutrient levels may have to be specifically requested. Soil nitrate profiles are used to determine nitrogen levels in the drier soils of the Prairies. A nitrogen test is now available in eastern Canada. Tissue analysis should be used if micronutrient deficiency is suspected. New tests are currently being developed which will help the organic farmer to gauge soil biological activity.

Some labs record results in ppm, others in lbs/acre. To convert ppm to lbs/acre, multiply by 2. If results indicate low nutrient levels, check to see if factors such as pH are limiting availability. Rotation plans may need to be modified to include more soil-building crops and more emphasis given to increasing organic matter. Very low levels suggest the need for soil amendments such as finely-ground rock powders or increased compost applications. Per cent base saturation of the exchangeable cations calcium, magnesium and potassium is given by some labs. It is claimed that this provides a guide for soil mineral balance with desired levels being potassium 2-7 per cent, magnesium 10-20 per cent and calcium 60-70 per cent. However, there is research that shows this is not appropriate in Ontario, especially where calcium content is naturally high. Organic matter levels of 4-5 per cent are considered good.

NITROGEN FIXATION

Nitrogen fixation is limited to prokaryotes. Some eubacteria and a few archebacteria can fix nitrogen–but no eukaryotic cells can do this. Some N-fixing bacteria are free-living whereas other form symbiotic associations with plants.

Free Living Nitrogen Fixing Bacteria:

- Obligate anaerobes, e.g., *Clostridium pasteurianum.*
- Facultative anaerobes, e.g., *Klebsiella,* a close relative of *E. coli.*
- Photosynthetic bacteria, e.g., *Rhodobacter.*
- Many cyanobacteria.
- Obligate aerobes such as *Azotobacter.*
- Some methanogens.

Since nitrogenase is inactivated by O_2, the fixation of N_2 must occur under conditions which are anaerobic at least locally. For anaerobes there is no problem. Facultative organisms such as purple photosynthetic bacteria or *Klesbsiella* fix N_2 only when anaerobic. Other organisms have protective mechanisms. In *Azotobacter,* an obligate aerobe, the O_2 concentration inside the cell is held down by partial uncoupling of a highly active respiratory chain. This wastes carbohydrate, but if growth is limited by absence of nitrogen compounds then this is justifiable. In cyanobacteria O_2 is actually generated by photosynthesis. Fixation of N_2 occurs in special cells known as heterocysts which do not photosynthesize but are devoted solely to N_2 fixation.

Symbiotic Nitrogen Fixing Bacteria

Symbiotic bacteria are protected from oxygen by inhabiting a plant host. Bacteria of the genus *Rhizobium* and *Bradyrhizobium* inhabit the root nodules of leguminous plants (e.g., peas, beans, clover, alfalfa, soybeans). Other symbiotic associations occur but are less important. *Anabaena azollae,* a nitrogen fixing cyanobacterium, lives in pores on the fronds of a water fern called *Azolla.*

This symbiotic partnership is used to enrich rice paddies with organic nitrogen. *Rhizobium* is also found free in the soil but only fixes N_2 when inside the root nodules of its host plant, in a strictly controlled microaerophilic environment. Oxygen is required to generate sufficient respiratory energy to drive N_2 fixation. But too much oxygen inactivates nitrogenase.

In root nodules the O_2 level is regulated by a special hemoglobin - leghaemoglobin. The globin protein is encoded by plant genes but the heme cofactor is made by the symbiotic bacteria. This is produced only when the plant is infected with *Rhizobium.*

The plant root cells convert sugar to organic acids which they supply to the bacteroids. In exchange, the plant receives amino-acids (rather than free ammonia). Specific strains of bacteria are found inhabiting specific plant species. For example, a carbohydrate binding protein (lectin) on the surface of root cells of clover (*Trifolium*) specifically binds to lipopolysaccharide of *Rhizobium trifolii* which contains 2-deoxyglucose. The bacteria then enter and produce cytokinins (a type of plant hormone) which promote the division of plant cells to form nodules. The bacteria lose their outer membranes and become irregular in shape–"bacteroids".

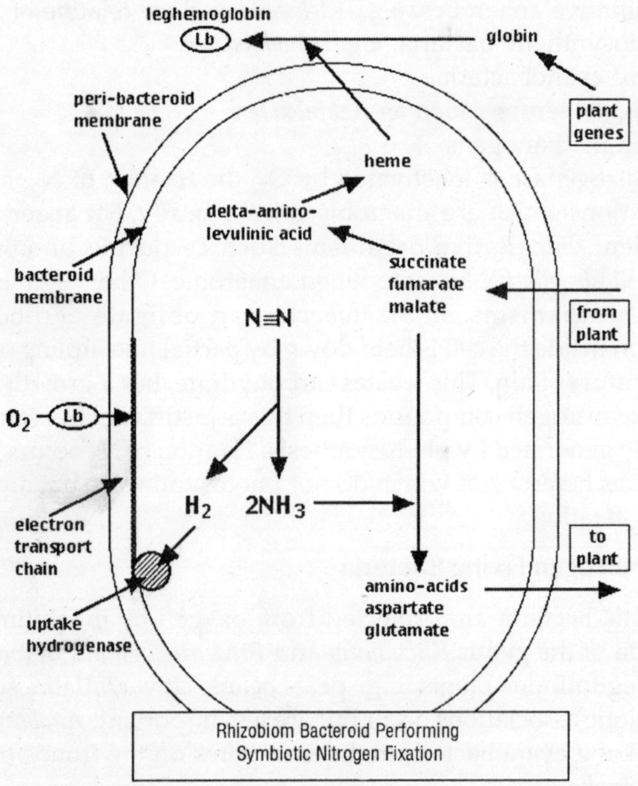

Fig:1. Rhizobium Bacteroid Performing Symbiotic Nitrogen Fixation

Structure and Operation of Nitrogenase

Nitrogenase contains the two proteins molybdoferredoxin and azoferredoxin. These must be supplied with reducing equivalents by other proteins that vary. Here we consider nitrogenase from *Klebsiella,* a close relative of *E. coli* where the accessory proteins are flavodoxin and pyruvate flavodoxin reductase. In most bacteria electrons are passed from NAD(P)H or pyruvate to ferredoxin, an FeS protein. If iron is in short supply ferredoxin is replaced by flavodoxin, a flavoprotein.

In *Klebsiella* there is no ferredoxin and flavodoxin (NifF protein) is used all the time. Azoferredoxin transfers electrons from reduced flavodoxin (or ferredoxin) to molybdoferredoxin.

Molybdoferredoxin is an alpha2/beta2 tetramer. The alpha and beta subunits are similar but distinct and are encoded by genes *nifK* and *nifD*. Each tetramer contains 2 Mo and several FeS groups. The molybdenum is part of a low molecular weight cofactor containing Mo bound to an Fe7S8 cluster and to homocitrate.

This MoFe cofactor is unique to nitrogen fixation and distinct from the Mo-pterin cofactor of other Mo proteins (e.g., nitrate reductase, xanthine oxidase).

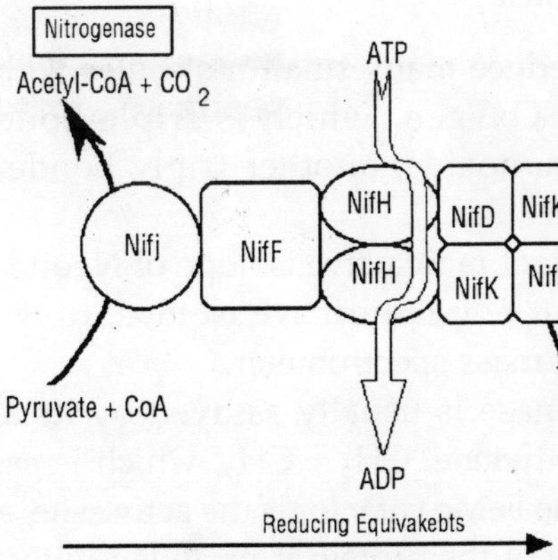

Fig:2. Nitrogenase

Azoferredoxin is a dimer of identical subunits encoded by *nifH* and contains a single Fe_4S_4 group per dimer. Azoferredoxin is modified by the NifM protein. Molybdoferredoxin from one genus can often interact with azoferredoxin from another genus to give active enzyme. These two proteins have several alternative names:

Molybdoferredoxin = component 1, MoFe protein, or "nitrogenase"

Azoferredoxin = component II, Fe protein, or nitrogenase reductase. Nitrogenase is not very fast (the turnover number is around 50 moles/ min per mole of Mo) and so about 2-5% of the total cell protein is nitrogenase. The reaction $N_2 + 3H_2 2NH_3$ actually releases energy. However, the activation energy needed to break the NºN triple bond is very high and in practice energy, as ATP, is consumed by NifH protein (azoferredoxin). If there is an excess of azoferredoxin then ATP tends to be wasted. In *Klebsiella nifHDK* form an operon that keeps the ratio of components constant.

The Nif (nitrogen fixation) proteins are often referred to by their gene names:

NifJ = pyruvate flavodoxin reductase
nifF = flavodoxin
nifH = azoferredoxin
nifM = processing of NifH protein
nifK, D = molybdoferredoxin
nifB, N, E, V, W, Z = MoFe cofactor synthesis

nifY = MoFe cofactor insertion
nifQ = molybdenum uptake
nifA,L, R= regulation
nifU, S = metal centre biosynthesis
nifX, T = function unknown (not necessary, at least under normal conditions)

Mechanism of Nitrogenase

Nitrogenase will reduce many small molecules with triple bonds in addition to nitrogen. Oxygen, which is triple-bonded inactivates nitrogenase. Carbon monoxide, another triply bonded molecule is a competitive inhibitor.

There is no convenient radioactive isotope of N_2 and the reduction of nitrogen-15, a heavy but non-radioactive isotope, to NH_3 is difficult to measure and requires a mass spectrometer.

In practice nitrogenase is usually assayed by its ability to reduce acetylene, CHºCH, to ethylene, $CH_2 = CH_2$, which is easily detected by gas chromatography. The FeMo cofactor is the active site, and it can reduce acetylene in the absence of the protein if provided with a good chemical reducing agent such as borohydride.

Nitrogenase Substrates Products
Nitrogen NºN $NH_3 + NH_3$
Acetylene HCºCH $H_2C = CH_2$
Cyanides HCºN $CH_4 + NH_3$
RCºN $RCH_3 + NH_3$
Azides H-N—N+ºN N_2 + NH3
Nitrous oxide NºN+-O- $N_2 + H_2O$
Isocyanides R-N+ºC- $RNH_2 + CH_4$

The overall Delta G for $N_2 + 3H_2 = 2NH_3$ is about -8 kcal/mole. However the first step, opening up the triple bond, is extremely unfavourable:

$N_2 + H_2 = N_2H_2$ Delta G° = +50 kcal/mole (approximately)
N_2 + 2e- + 2H+ = N_2H_2 Eo = –1200 mV (approximately)

Thus nitrogenase has to carry out a single step which needs a reductant with a redox potential of –1200 mV. Alterations in solvation, local pH, etc., could bring this down to about –1000 mV but even so this is much more negative than any other biological redox potential.

The redox potential of azoferredoxin is –290 mV. [When ATP binds to this protein its redox potential is lowered to –400 mV.] The ATP must be hydrolysed for reduced azoferredoxin to reduce molybdoferredoxin. Two ATP are hydrolysed per electron transferred or 4ATP/2e. 4ATP yields approximately –30 kcal which is equivalent to 750 mV per pair of electrons. Adding this 750 mV to the Eo of azoferredoxin (–290 mV) just over 1000 mV negative – the correct value. This suggests that ATP is used to generate reducing power.

The mechanism is unknown, but remember that cells convert reducing power to ATP during respiration. N_2 is reduced at the MoFe cofactor site on the molybdofer-redoxin. The intermediates N_2H_2 and N_2H_4 (hydrazine) are assumed to exist. Although N_2H_4 has been detected, N_2H_2 is very unstable and tends to decompose back to $N_2 + H_2$.

Hydrogen is always produced when nitrogenase reduces N_2 to NH_3. There are two views on this. The first is that this is a side reaction nitrogenase is such a powerful reductant that conversion of H_2O to H_2 inevitably occurs.

The second view is that under optimum conditions one H_2 is evolved per N_2 fixed suggesting that H_2 evolution is an integral part of the enzyme mechanism. Furthermore, reduction of acetylene to ethylene is not accompanied by H_2 evolution. Since the Eo for acetylene is +320 mV it is possible that nitrogenase reduces acetylene when only partly activated and H_2 evolution is not necessary in this case.

Proposed Steps in Nitrogenase Mechanism

The mechanism is largely based on work with non-protein MoFe complexes, some of which will fix N_2 chemically (but very inefficiently).

- Mo in active site is reduced from Mo^{6+} to Mo^{5+} to Mo^{4+} by sequential electron transfer from azoferredoxin.
- Semi-activated nitrogenase can reduce easy substrates such as acetylene.
- Further transfer of two electrons activates the Fe of the MoFe cofactor in the active site, which carries 2H.
- N_2 binds end on to the FeH_2 complex and releases H_2.
- The bound N_2 is reduced to HN=NH by sideways transfer of 2e- (plus $2H^+$) from the active site Mo^{4+}.
- Conversion of N_2H_2 to $2NH_3$ requires two further $2e^-$ steps, but partial activation of the enzyme is sufficient (i.e., ATP is no longer needed to hype up the redox potential) since only step (e) requires extreme reducing power.

This mechanism also explains why acetylene, C_2H_2, is a non-competitive inhibitor of N_2 fixation. Acetylene reduction discharges nitrogenase before it ever reaches full activation.

Although N_2 fixation wastes reducing power when H_2 is evolved, most N_2 fixing bacteria contain hydrogenase which uses gaseous H_2 to reduce NAD(P). Hence they recycle the hydrogen at least partly.

Nitrogenase Mechanism

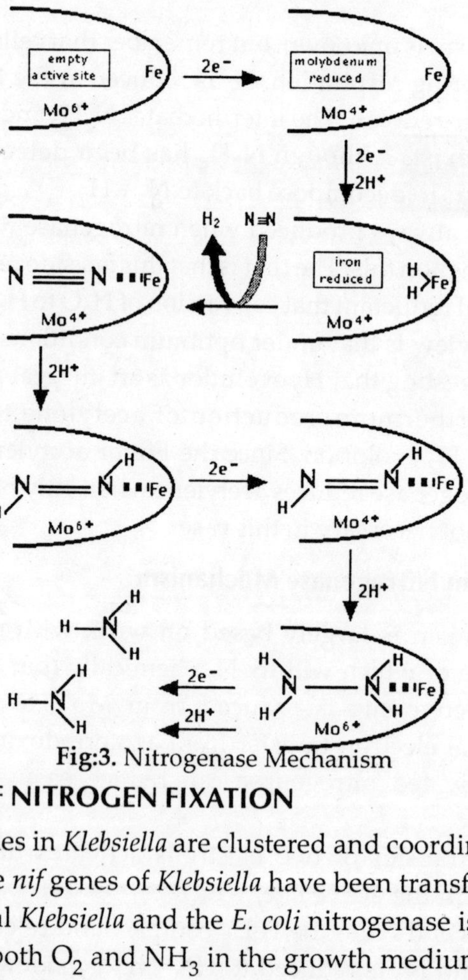

Fig:3. Nitrogenase Mechanism

REGULATION OF NITROGEN FIXATION

All the *nif* genes in *Klebsiella* are clustered and coordinately regulated. *E. coli* to which the *nif* genes of *Klebsiella* have been transferred can fix N_2. In both the original *Klebsiella* and the *E. coli* nitrogenase is expressed only in the absence of both O_2 and NH_3 in the growth medium. Other organic N-sources will also repress nitrogenase. The better the N-source the greater the repression. The *nif* genes are regulated by the *nifLA* operon. The nitrogen regulators NtrC (= GlnG), and NtrB determine whether or not the *nifLA* operon is expressed (depending on the presence of ammonia or organic nitrogen). In the absence of ammonia or organic nitrogen the NtrC protein is phosphorylated by the NtrB protein. NtrC-P then binds to the upstream region of the *nifLA* operon and activates transcription.

NtrA (= GlnF = RpoN = s54) is the nitrogen sigma factor, which is needed for expression of the *nifLA* operon and the *nif* structural genes.

NtrA is an alternative sigma factor used by RNA polymerase to recognise many genes involved in nitrogen metabolism which are not recognised by the standard sigma factor. The *nifA* gene encodes a protein required for switching on all of the *nif* genes except the regulatory genes *nifLA* themselves. if NifA protein is made, its function is to activate the other *nif* genes. The *nifL* gene is required for O_2 repression. In the absence of NifL protein, nitrogenase is made in the presence of O_2 (but is inactivated by O_2). When oxygen is present, the *NifL* protein binds to *NifA* and prevents it from activating the other *nif* genes.

ALTERNATIVE NITROGEN FIXATION SYSTEMS

When Mo is absent some N-fixing bacteria, such as *Azotobacter*, make an alternative nitrogenase in which vanadium is used instead of Mo. This is encoded by a duplicate set of *vnf* genes which make VFe cofactor as well as the corresponding nitrogenase proteins. Mo, if available represses the *vnf* system which is less efficient. If vanadium is also absent *Azotobacter* can make a third nitrogenase which uses only iron – the even less efficient *anf* system. The sequences of the *nif*, *vnf* and *anf* genes are very similar.

REGULATION OF NITROGEN FIXATION

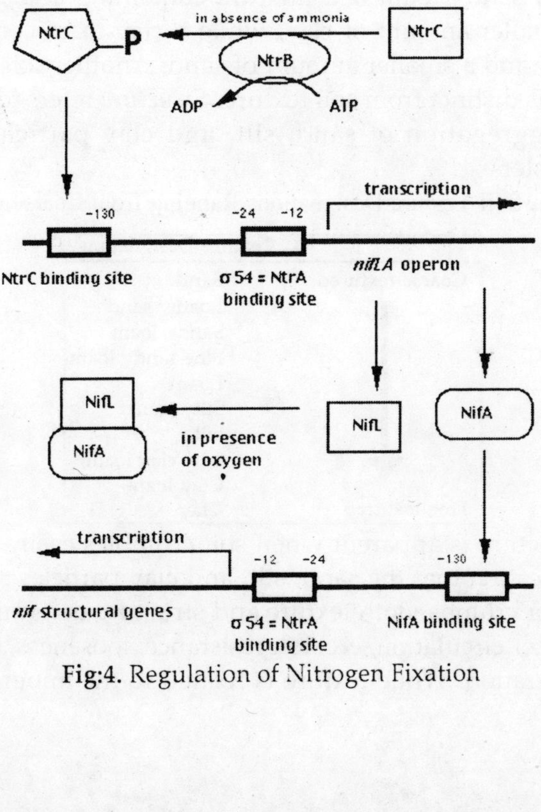

Fig:4. Regulation of Nitrogen Fixation

SOIL MICROORGANISMS : THE LIVING SOIL

Textile and Structure

Soils are made up of four basic components: minerals, air, water, and organic matter. In most soils, minerals represent around 45% of the total volume, water and air about 25% each, and organic matter from 2% to 5%. The mineral portion consists of three distinct particle sizes classified as sand, silt, or clay. Sand is the largest particle that can be considered soil. Sand is largely the mineral quartz, though other minerals are also present. Quartz contains no plant nutrients, and sand cannot hold nutrients, they leach out easily with rainfall. Silt particles are much smaller than sand, but like sand, silt is mostly quartz.

The smallest of all the soil particles is clay. Clays are quite different from sand or silt, and most types of clay contain appreciable amounts of plant nutrients. Clay has a large surface area resulting from the plate-like shape of the individual particles. Sandy soils are less productive than silts, while soils containing clay are the most productive and use fertilizers most effectively.

Soil texture refers to the relative proportions of sand, silt, and clay. A loam soil contains these three types of soil particles in roughly equal proportions. A sandy loam is a mixture containing a larger amount of sand and a smaller amount of clay, while a clay loam contains a larger amount of clay and a smaller amount of sand. Another soil characteristic, soil structure, is distinct from soil texture. *Structure* refers to the clumping together or aggregation of sand, silt, and clay particles into larger secondary clusters.

Table:1. Soil Texture Designations Ranging from Coarse to Fine.

	Texture Designation
Coarse-textured	Sand
↓	Loamy sand
	Sandy loam
	Fine sandy loam
	Loam
	Silty loam
	Silt
	Silty clay loam
	Clay loam
Fine-textured	Clay

Good structure is apparent when soil crumbles easily in your hand. This is an indication that the sand, silt, and clay particles are aggregated into granules or crumbs. Both texture and structure determine pore space for air and water circulation, erosion resistance, looseness, ease of tillage, and root penetration. While texture is related to the minerals in soil and does not change with agricultural activities, structure can be improved or destroyed readily by choice and timing of farm practices.

The Living Soil: The Importance of Soil Organisms

An acre of living topsoil contains approximately 900 pounds of earthworms, 2,400 pounds of fungi, 1,500 pounds of bacteria, 133 pounds of protozoa, 890 pounds of arthropods and algae, and even small mammals in some cases. Therefore, the soil can be viewed as a living community rather than an inert body. Soil organic matter also contains dead organisms, plant matter, and other organic materials in various phases of decomposition. Humus, the dark-coloured organic material in the final stages of decomposition, is relatively stable.

Both organic matter and humus serve as reservoirs of plant nutrients; they also help to build soil structure and provide other benefits. The type of healthy living soil required to support humans now and far into the future will be balanced in nutrients and high in humus, with a broad diversity of soil organisms. It will produce healthy plants with minimal weed, disease, and insect pressure. To accomplish this, we need to work *with* the natural processes and optimise their functions to sustain our farms. Considering the natural landscape, you might wonder how native prairies and forests function in the absence of tillage and fertilizers. These soils are tilled by soil organisms, not by machinery.

They are fertilized too, but the fertility is used again and again and never leaves the site. Native soils are covered with a layer of plant litter and/or growing plants throughout the year. Beneath the surface litter, a rich complexity of soil organisms decompose plant residue and dead roots, then release their stored nutrients slowly over time. In fact, topsoil is the most biologically diverse part of the earth. Soil-dwelling organisms release bound-up minerals, converting them into plant-available forms that are then taken up by the plants growing on the site.

The organisms recycle nutrients again and again with the death and decay of each new generation of plants. There are many different types of creatures that live on or in the topsoil. Each has a role to play. These organisms will work for the farmer's benefit if we simply manage for their survival. Consequently, we may refer to them as soil livestock. While a great variety of organisms contribute to soil fertility, earthworms, arthropods, and the various microorganisms merit particular attention.

Earth Worms

Earthworm burrows enhance water infiltration and soil aeration. Fields that are tilled by earthworm tunneling can absorb water at a rate 4 to 10 times that of fields lacking worm tunnels. This reduces water runoff, recharges groundwater, and helps store more soil water for dry spells.

Vertical earthworm burrows pipe air deeper into the soil, stimulating microbial nutrient cycling at those deeper levels. When earthworms are present in high numbers, the tillage provided by their burrows can replace some expensive tillage work done by machinery.

Worms eat dead plant material left on top of soil and redistribute the organic matter and nutrients throughout the topsoil layer. Nutrient- rich organic compounds line their tunnels, which may remain in place for years if not disturbed. During droughts these tunnels allow for deep plant root penetration into subsoil regions of higher moisture content. In addition to organic matter, worms also consume soil and soil microbes. The soil clusters they expel from their digestive tracts are known as *worm casts* or *castings*. These range from the size of a mustard seed to that of a sorghum seed, depending on the size of the worm.

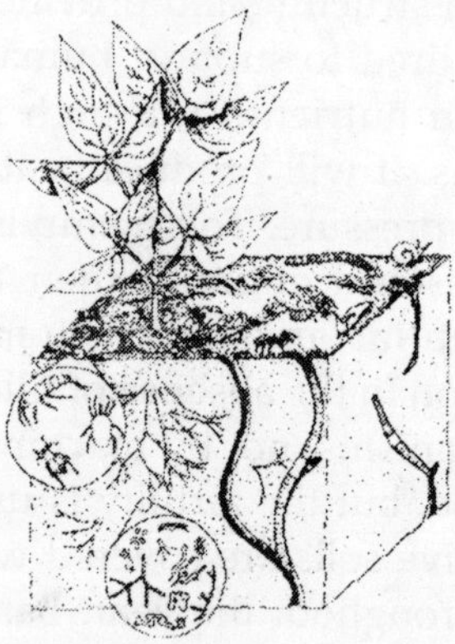

Fig:5. The Soil is Teeming with Organisms that Cycle Nutrients from Soil to Plants and Back Again

The soluble nutrient content of worm casts is considerably higher than that of the original soil. A good population of earthworms can process 20,000 pounds of topsoil per year, with turnover rates as high as 200 tons per acre having been reported in some exceptional cases. Earthworms also secrete a plant growth stimulant. Reported increases in plant growth following earthworm activity may be partially attributed to this substance, not just to improved soil quality.

Table:2. Selected Nutrient Analyses of Worm Casts Compared to Those of the Surrounding Soil

Nutrient	*Worm Casts*	*Soil*
	Lbs/ac	Lbs/ac
Carbon	171,000	78,500
Nitrogen	10,700	7,000
Phosphorus	280	40
Potassium	900	140

Earthworms thrive where there is no tillage. Generally, the less tillage the better, and the shallower the tillage the better. Worm numbers can be reduced by as much as 90% by deep and frequent tillage. Tillage reduces earthworm populations by drying the soil, burying the plant residue they feed on, and making the soil more likely to freeze. Tillage also destroys vertical worm burrows and can kill and cut up the worms themselves. Worms are dormant in the hot part of the summer and in the cold of winter.

Young worms emerge in spring and fall. they are most active just when farmers are likely to be tilling soil.

Table:3. Effect of Crop Management on Earthworm Populations

Crop	*Management*	*Worms/Foot^2*
Corn	Plow	1
Corn	No-Till	2
Soybean	Plow	6
Soybean	No-Till	14
Bluegrass/clover	....	39
Dairy pasture	....	33

As a rule, earthworm numbers can be increased by reducing or eliminating tillage (especially fall tillage), not using a moldboard plow, reducing residue particle size (using a straw chopper on the combine), adding animal manure, and growing green manure crops. It is beneficial to leave as much surface residue as possible year-round.

Cropping systems that typically have the most earthworms are (in descending order) perennial cool-season grass grazed rotationally, warm season perennial grass grazed rotationally, and annual croplands using no-till. Ridge-till and strip tillage will generally have more earthworms than clean tillage involving plowing and disking. Cool season grass rotationally grazed is highest because it provides an undisturbed (no-tillage) environment plus abundant organic matter from the grass roots and fallen grass litter. Generally speaking, worms want their food on top, and they want to be left alone.

Earthworms prefer a near-neutral soil pH, moist soil conditions, and plenty of plant residue on the soil surface. They are sensitive to certain pesticides and some incorporated fertilizers. Carbamate insecticides, including Furadan, Sevin, and Temik, are harmful to earthworms, notes worm biologist Clive Edwards of Ohio State University. Some insecticides in the organophosphate family are mildly toxic to earthworms, while synthetic pyrethroids are harmless to them.

Most herbicides have little effect on worms except for the triazines, such as Atrazine, which are moderately toxic. Also, anhydrous ammonia kills earthworms in the injection zone because it dries the soil and temporarily increases the pH there. High rates of ammonium-based fertilizers are also harmful.

Arthropods

In addition to earthworms, there are many other species of soil organisms that can be seen by the naked eye. Among them are sowbugs, millipedes, centipedes, slugs, snails, and springtails. These are the primary decomposers. Their role is to eat and shred the large particles of plant and animal residues. Some bury residue, bringing it into contact with other soil organisms that further decompose it. Some members of this group prey on smaller soil organisms. The springtails are small insects that eat mostly fungi. Their waste is rich in plant nutrients released after other fungi and bacteria decompose it. Also of interest are dung beetles, which play a valuable role in recycling manure and reducing livestock intestinal parasites and flies.

Bacteria

Bacteria are the most numerous type of soil organism: every gramme of soil contains at least a million of these tiny one-celled organisms. There are many different species of bacteria, each with its own role in soil environment. One of the major benefits bacteria provide for plants is in making nutrients available to them.

Some species release nitrogen, sulfur, phosphorus, and trace elements from organic matter. Others break down soil minerals, releasing potassium, phosphorus, magnesium, calcium, and iron. Still other species make and release plant growth hormones, which stimulate root growth. Several species of bacteria transform nitrogen from a gas in the air to forms available for plant use, and from these forms back to a gas again. A few species of bacteria fix nitrogen in the roots of legumes, while others fix nitrogen independently of plant association. Bacteria are responsible for converting nitrogen from ammonium to nitrate and back again, depending on certain soil conditions. Other benefits to plants provided by various species of bacteria include increasing the solubility of nutrients, improving soil structure, fighting root diseases, and detoxifying soil.

Fungi

Fungi come in many different species, sizes, and shapes in soil. Some species appear as threadlike colonies, while others are one-celled yeasts. Slime molds and mushrooms are also fungi. Many fungi aid plants by breaking down organic matter or by releasing nutrients from soil minerals. Fungi are generally quick to colonize larger pieces of organic matter and begin the decomposition process. Some fungi produce plant hormones, while others produce antibiotics including penicillin.

There are even species of fungi that trap harmful plant-parasitic nematodes. The mycorrhizae are fungi that live either on or in plant roots and act to extend the reach of root hairs into the soil. Mycorrhizae increase the uptake of water and nutrients, especially phosphorus. They are particularly important in degraded or less fertile soils. Roots colonised by mycorrhizae are less likely to be penetrated by root-feeding nematodes, since the pest cannot pierce the thick fungal network. Mycorrhizae also produce hormones and antibiotics that enhance root growth and provide disease suppression. The fungi benefit by taking nutrients and carbohydrates from the plant roots they live in.

Actinomycetes

Actinomycetes are threadlike bacteria that look like fungi. While not as numerous as bacteria, they too perform vital roles in the soil. Like the bacteria, they help decompose organic matter into humus, releasing nutrients. They also produce antibiotics to fight diseases of roots. Many of these same antibiotics are used to treat human diseases. Actinomycetes are responsible for the sweet, earthy smell noticed whenever a biologically active soil is tilled.

Algae

Many different species of algae live in the upper half-inch of the soil. Unlike most other soil organisms, algae produce their own food through photosynthesis. They appear as a greenish film on the soil surface following a saturating rain. Algae improve soil structure by producing slimy substances that glue soil together into water-stable aggregates. Some species of algae (the blue-greens) can fix their own nitrogen, some of which is later released to plant roots.

Protozoa

Protozoa are free-living microorganisms that crawl or swim in the water between soil particles. Many soil protozoa are predatory, eating other microbes. One of the most common is an amoeba that eats bacteria. By eating and digesting bacteria, protozoa speed up the cycling of nitrogen from the bacteria, making it more available to plants.

Nematodes

Nematodes are abundant in most soils, and only a few species are harmful to plants. The harmless species eat decaying plant litter, bacteria, fungi, algae, protozoa, and other nematodes. Like other soil predators, nematodes speed the rate of nutrient cycling.

Soil Organisms and Soil Quality

All these organisms from the tiny bacteria up to the large earthworms and insects, interact with one another in a multitude of ways in the soil ecosystem. Organisms not directly involved in decomposing plant wastes may feed on each other or each other's waste products or the other substances they release. Among the substances released by the various microbes are vitamins, amino acids, sugars, antibiotics, gums, and waxes. Roots can also release into the soil various substances that stimulate soil microbes. These substances serve as food for select organisms.

Table:4. Weights of Soil Organisms in the Top 7 inches of Fertile Soil

Organism	*Pounds of Liverweight/Acre*
Bacteria	1000
Actinomycetes	1000
Molds	2000
Algae	100
Protozoa	200
Nematodes	50
Insects	100
Worms	1000
Plant Roots	2000

Some scientists and practitioners theorise that plants use this means to stimulate the specific population of microorganisms capable of releasing or otherwise producing the kind of nutrition needed by the plants. Research on life in the soil has determined that there are ideal ratios for certain key organisms in highly productive soils.

The Soil Foodweb Lab, located in Oregon, tests soils and makes fertility recommendations that are based on this understanding. Their goal is to alter the makeup of the soil microbial community so it resembles that of a highly fertile and productive soil. There are several different ways to accomplish this goal, depending on the situation.

ORGANIC MATTER, HUMUS AND THE SOIL FOODWEB

Understanding the role that soil organisms play is critical to sustainable soil management. Based on that understanding, focus can be directed toward strategies that build both the numbers and the diversity of soil organisms. Like cattle and other farm animals, soil livestock require proper feed.

That feed comes in the form of organic matter. *Organic matter* and *humus* are terms that describe somewhat different but related things. Organic matter refers to the fraction of soil that is composed of both living organisms and once-living residues in various stages of decomposition. Humus is only a small portion of the organic matter. It is the end product of organic matter decomposition and is relatively stable. Further decomposition of humus occurs very slowly in both agricultural and natural settings.

In natural systems, a balance is reached between the amount of humus formation and the amount of humus decay. This balance also occurs in most agricultural soils, but often at a much lower level of soil humus.

Humus contributes to well-structured soil that, in turn, produces high-quality plants. It is clear that management of organic matter and humus is essential to sustaining the whole soil ecosystem. The benefits of a topsoil rich in organic matter and humus are many.

They include rapid decomposition of crop residues, granulation of soil into water-stable aggregates, decreased crusting and clodding, improved internal drainage, better water infiltration, and increased water and nutrient holding capacity. Improvements in the soil's physical structure facilitate easier tillage, increased water storage capacity, reduced erosion, better formation and harvesting of root crops, and deeper, more prolific plant root systems. Soil organic matter can be compared to a bank account for plant nutrients.

Soil containing 4% organic matter in the top seven inches has 80,000 pounds of organic matter per acre. That 80,000 pounds of organic matter will contain about 5.25% nitrogen, amounting to 4,200 pounds of nitrogen per acre. Assuming a 5% release rate during the growing season, the organic matter could supply 210 pounds of nitrogen to a crop. However, if the organic matter is allowed to degrade and lose nitrogen, purchased fertilizer will be necessary to prop up crop yields. All the soil organisms mentioned previously, except algae, depend on organic matter as their food source.

Therefore, to maintain their populations, organic matter must be renewed from plants growing on the soil, or from animal manure, compost, or other materials imported from off site. When soil livestock are fed, fertility is built up in the soil, and the soil will feed the plants. Ultimately, building organic matter and humus levels in the soil is a matter of managing the soil's living organisms, something akin to wildlife management or animal husbandry. This entails working to maintain favourable conditions of moisture, temperature, nutrients, pH, and aeration. It also involves providing a steady food source of raw organic material.

SOIL TILTH AND ORGANIC MATTER

A soil that drains well, does not crust, takes in water rapidly, and does not make clods is said to have good tilth. Tilth is the physical condition of the soil as it relates to tillage ease, seedbed quality, easy seedling emergence, and deep root penetration.

Good tilth is dependent on aggregation, the process whereby individual soil particles are joined into clusters or aggregates. Aggregates form in soils when individual soil particles are oriented and brought together through the physical forces of wetting and drying or freezing and thawing. Weak electrical forces from calcium and magnesium hold soil particles together when soil dries. When these aggregates become wet again, however, their stability is challenged, and they may break apart. Aggregates can also be held together by plant roots, earthworm activity, and by glue-like products produced by soil microorganisms. Earthworm-created aggregates are stable once they come out of the worm.

An aggregate formed by physical forces can be bound together by fine root hairs or threads produced by fungi. Aggregates can also become stabilised (remain intact when wet) through the by-products of organic matter decomposition by fungi and bacteria, chiefly gums, waxes, and other glue-like substances. These by-products cement the soil particles together, forming water-stable aggregates.

The aggregate is then strong enough to hold together when wet, hence the term water-stable. USDA soil microbiologist Sara Wright named the glue that holds aggregates together, glomalin, after the Glomales group of common root-dwelling fungi. These fungi secrete a gooey protein known as glomalin through their hair-like filaments, or hyphae. When Wright measured glomalin in soil aggregates she found levels as high as 2% of their total weight in eastern US soils. Soil aggregates from the West and Midwest had lower levels of glomalin. She found that tillage tends to lower glomalin levels.

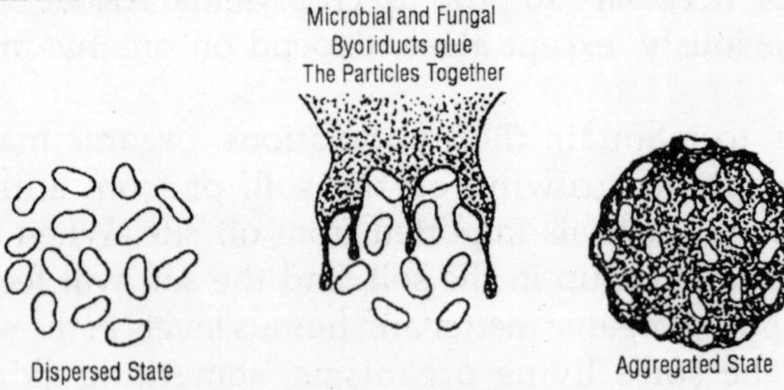

Fig:6. Macrobial Byproducts Glue Soil Particles into Water-stable Aggregates

Glomalin levels and aggregation were higher in no-till corn plots than in tilled plots. Wright has a brochure describing glomalin and how it benefits soil, entitled *Glomalin, a Manageable Soil Glue*. To order this brochure.

A well-aggregated soil allows for increased water entry, increased air flow, and increased water-holding capacity. Plant roots occupy a larger volume of well-aggregated soil, high in organic matter, as compared to a finely pulverised and dispersed soil, low in organic matter. Roots, earthworms, and soil arthropods can pass more easily through a well-aggregated soil.

Aggregated soils also prevent crusting of the soil surface. Finally, well-aggregated soils are more erosion resistant, because aggregates are much heavier than their particle components. For a good example of the effect of organic matter additions on aggregation.

Table:5. Water entry into the Soil after 1 hour

Manure Rate (tons/acre)	*Inches of Water*
0	1.2
8	1.9
16	2.7

The opposite of aggregation is dispersion. In a dispersed soil, each individual soil particle is free to blow away with the wind or wash away with overland flow of water. Clay soils with poor aggregation tend to be sticky when wet, and cloddy when dry. If the clay particles in these soils can be aggregated together, better aeration and water infiltration will result. Sandy soils can benefit from aggregation by having a small amount of dispersed clay that tends to stick between the sand particles and slow the downward movement of water.

Crusting is a common problem on soils that are poorly aggregated. Crusting results chiefly from the impact of falling raindrops. Rainfall causes clay particles on the soil surface to disperse and clog the pores immediately beneath the surface. Following drying, a sealed soil surface results in which most of the pore space has been drastically reduced due to clogging from dispersed clay particles. Subsequent rainfall is much more likely to run off than to flow into the soil.

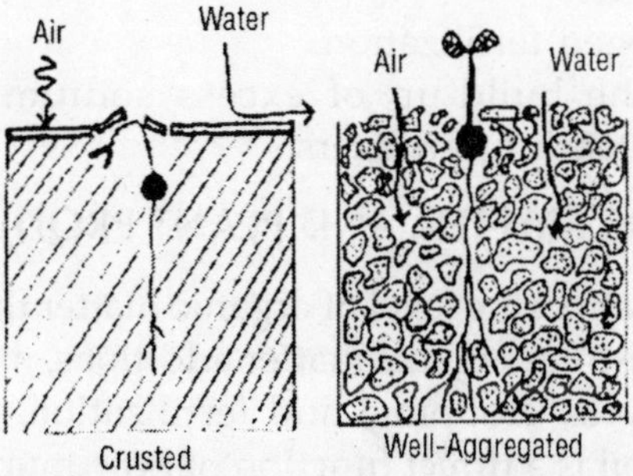

Fig:7. Effects of Aggregation on Water and Air Entry into the Soil

Since raindrops start crusting, any management practices that protect soil from their impact will decrease crusting and increase water flow into soil. Mulches and cover crops serve this purpose well, as do no-till practices, which allow the accumulation of surface residue. Also, a well-aggregated soil will resist crusting because the water-stable aggregates are less likely to break apart when a raindrop hits them.

Long-term grass production produces the bestaggregated soils. A grass sod extends a mass of fine roots throughout the topsoil, contributing to the physical processes that help form aggregates. Roots continually remove water from soil microsites, providing local wetting and drying effects that promote aggregation. Fine root hairs also bind soil aggregates together. Roots also produce food for soil microorganisms and earthworms, which in turn generate compounds that bind soil particles into water-stable aggregates.

In addition, perennial grass sods provide protection from raindrops and erosion. Thus, a perennial cover creates a combination of conditions optimal for the creation and maintenance of well-aggregated soil. Conversely, cropping sequences that involve annual plants and extensive cultivation provide less vegetative cover and organic matter, and usually result in a rapid decline in soil aggregation. Farming practices can be geared to conserve and promote soil aggregation. Because the binding substances are themselves susceptible to microbial degradation, organic matter needs to be replenished to maintain microbial populations and overall aggregated soil status. Practices should conserve aggregates once they are formed, by minimising factors that degrade and destroy aggregation.

Some factors that destroy or degrade soil aggregates are:

- Bare soil surface exposed to the impact of raindrops.
- Removal of organic matter through crop production and harvest without return of organic matter to soil.
- Excessive tillage, working the soil when it is too wet or too dry.
- Use of anhydrous ammonia, which speeds up decomposition of organic matter.
- Excess nitrogen fertilization.
- Allowing the build-up of excess sodium from irrigation or sodium-containing fertilizers.

TILLAGE, ORGANIC MATTER, AND PLANT PRODUCTIVITY

Several factors affect the level of organic matter that can be maintained in a soil. Among these are organic matter additions, moisture, temperature, tillage, nitrogen levels, cropping, and fertilization. The level of organic matter present in soil is a direct function of how much organic material is being produced or added to the soil versus the rate of decomposition. Achieving this balance entails slowing the speed of organic matter decomposition, while increasing the supply of organic materials produced on site and/or added from off site.

Moisture and temperature also profoundly affect soil organic matter levels. High rainfall and temperature promote rapid plant growth, but these conditions are also favourable to rapid organic matter decomposition and loss. Low rainfall or low temperatures slow both plant growth and organic matter decomposition.

The native Midwest Prairie soils originally had a high amount of organic matter from the continuous growth and decomposition of perennial grasses, combined with a moderate temperature that did not allow for rapid decomposition of organic matter. Moist and hot tropical areas may appear lush because of rapid plant growth, but soils in these areas are low in nutrients. Rapid decomposition of organic matter returns nutrients back to soil, where they are almost immediately taken up by rapidly growing plants. Tillage can be beneficial or harmful to a biologically active soil, depending on what type of tillage is used and when it is done.

Tillage affects both erosion rates and soil organic matter decomposition rates. Tillage can reduce the organic matter level in croplands below 1%, rendering them biologically dead. Clean tillage involving moldboard plowing and disking breaks down soil aggregates and leaves soil prone to erosion from wind and water. The moldboard plow can bury crop residue and topsoil to a depth of 14 inches. At this depth, the oxygen level in soil is so low that decomposition cannot proceed adequately. Surface-dwelling decomposer organisms suddenly find themselves suffocated and soon die. Crop residues that were originally on the surface but now have been turned under will putrefy in the oxygendeprived zone.

This rotting activity may give a putrid smell to the soil. Furthermore, the top few inches of the field are now often covered with subsoil having very little organic matter content and, therefore, limited ability to support productive crop growth. The topsoil is where the biological activity happens, it's where the oxygen is. That's why a fence post rots off at the surface. In terms of organic matter, tillage is similar to opening the air vents on a wood-burning stove; adding organic matter is like adding wood to the stove.

Ideally, organic matter decomposition should proceed as an efficient burn of the wood to release nutrients and carbohydrates to soil organisms and create stable humus. Shallow tillage incorporates residue and speeds the decomposition of organic matter by adding oxygen that microbes need to become more active. In cold climates with a long dormant season, light tillage of a heavy residue may be beneficial; in warmer climates it is hard enough to maintain organic matter levels without any tillage. The plow lays soil up on its side, increasing the surface area exposed to oxygen.

The other three types of tillage are intermediate in their ability to foster organic matter decomposition. Oxygen is the key factor here. The moldboard plow increases soil surface area, allowing more air into soil and speeding the decomposition rate. With the moldboard plow, more than the entire organic matter contribution from the wheat straw is gone within only 19 days following tillage. Finally, the passage of heavy equipment increases compaction in the wheel tracks, and some tillage implements themselves compact soil further, removing oxygen and increasing the chance that deeply buried residues will putrefy.

Tillage also reduces the rate of water entry into the soil by removal of ground cover and destruction of aggregates, resulting in compaction and crusting. Table 6 shows three different tillage methods and how they affect water entry into the soil. Notice the direct relationship between tillage type, ground cover, and water infiltration.

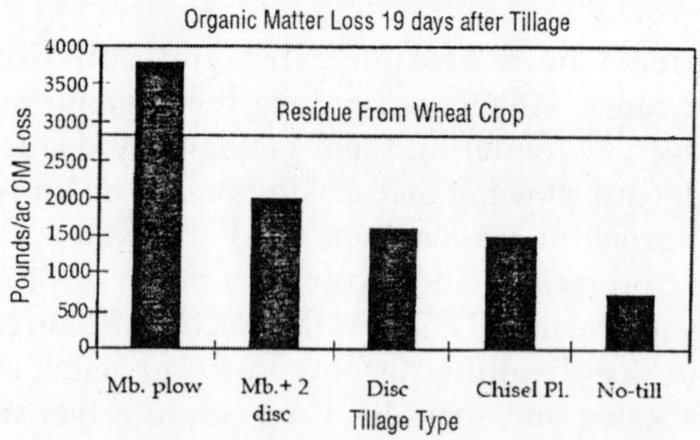

Fig:8. Organic Matter Loss 19 Days after Tillage

No-till has more than three times the water infiltration of the moldboard-plowed soil. Additionally, no-till fields will have higher aggregation from the organic matter decomposition on site. The surface mulch typical of no-till fields acts as a protective skin for soil. This soil skin reduces the impact of raindrops and buffers soil from temperature extremes as well as reducing water evaporation.

Table:6. Tillage Effects on Water Infiltration and Ground Cover

	Water Infiltration	*Ground Cover*
mm/minute per cent		
No-till	2.7	48
Chisel Plow	1.3	27
Moldboard Plow	0.8	12

Both no-till and reduced-tillage systems provide benefits to the soil. The advantages of a no-till system include superior soil conservation, moisture conservation, reduced water runoff, long-term buildup of organic matter, and increased water infiltration. A soil managed without tillage relies on soil organisms to take over the job of plant residue incorporation formerly done by tillage. On the down side, no-till can foster a reliance on herbicides to control weeds and can lead to soil compaction from the traffic of heavy equipment.

Pioneering development work on chemical-free no-till farming is proceeding at several research stations and farms in the eastern US Pennsylvania farmer Steve Groff has been farming notill with minimal or no herbicides for several years. Groff grows cover crops extensively in his fields, rolling them down in the spring using a 10-foot rolling stalk chopper. This rolling chopper kills the rye or vetch cover crop and creates a nice no-till mulch into which he plants a variety of vegetable and grain crops. After several years of no-till production, his soils are mellow and easy to plant into. Groff farms 175 acres of vegetables, alfalfa, and grain crops on his Cedar Meadow Farm.

Other conservation tillage systems include ridge tillage, minimum tillage, zone tillage, and reduced tillage, each possessing some of the advantages of both conventional till and no-till. These systems represent intermediate tillage systems, allowing more flexibility than either a notill or conventional till system might.

They are more beneficial to soil organisms than a conventional clean-tillage system of moldboard plowing and disking. Adding manure and compost is a recognised means for improving soil organic matter and humus levels. In their absence, perennial grass is the only crop that can regenerate and incréase soil humus. Cool-season grasses build soil organic matter faster than warm-season grasses because they are growing much longer during a given year.

When soil is warm enough for soil organisms to decompose organic matter, cool-season grass is growing. While growing, it is producing organic matter and cycling minerals from the decomposing organic matter in soil. In other words, there is a net gain of organic matter because the cool-season grass is producing organic matter faster than it is being used up. With warm-season grasses, organic matter production during the growing season can be slowed during the long dormant season from fall through early spring.

During the beginning and end of this dormant period, soil is still biologically active, yet no grass growth is proceeding. Some net accumulation of organic matter can occur under warmseason grasses, however. In a Texas study, switchgrass (a warm-season grass) grown for four years increased soil carbon content from 1.1% to 1.5% in the top 12 inches of soil. In hot and moist regions, a cropping rotation that includes several years of pasture will be most beneficial.

EFFECT OF NITROGEN ON ORGANIC MATTER

Excessive nitrogen applications stimulate increased microbial activity, which in turn speeds organic matter decomposition. The extra nitrogen narrows the ratio of carbon to nitrogen in the soil. Native or uncultivated soils have approximately 12 parts of carbon to each part of nitrogen, or a C:N ratio of 12:1. At this ratio, populations of decay bacteria are kept at a stable level since additional growth in their population is limited by a lack of nitrogen. When large amounts of inorganic nitrogen are added, the C:N ratio is reduced, which allows the populations of decay organisms to explode as they decompose more organic matter with the now abundant nitrogen. While soil bacteria can efficiently use moderate applications of inorganic nitrogen accompanied by organic amendments (carbon), excess nitrogen results in decomposition of existing organic matter at a rapid rate. Eventually, soil carbon content may be reduced to a level where the bacterial populations are on a starvation diet.

With little carbon available, bacterial populations shrink, and less of the free soil nitrogen is absorbed. Thereafter, applied nitrogen, rather than being cycled through microbial organisms and re-released to plants slowly over time, becomes subject to leaching. This can greatly reduce the efficiency of fertilization and lead to environmental problems. To minimise the fast decomposition of soil organic matter, carbon should be added with nitrogen.

Typical carbon sources, such as green manures, animal manure, and compost, serve this purpose well. Amendments containing too high a carbon to nitrogen ratio (25:1 or more) can tip the balance the other way, resulting in nitrogen being tied up in an unavailable form. Soil organisms consume all the nitrogen in an effort to decompose the abundant carbon; tied up in the soil organisms, nitrogen remains unavailable for plant uptake. As soon as a soil microorganism dies and decomposes, its nitrogen is consumed by another soil organism, until the balance between carbon and nitrogen is achieved again.

FERTILIZER AND BIOLOGICALLY ACTIVE SOILS

What are the soil mineral conditions that foster biologically active soils? Soil nutrients into a balance so that none were in excess or deficient. Albrecht's theory (also called base-saturation theory) is used to guide lime and fertilizer application by measuring and evaluating the ratios of positively charged nutrients (bases) held in soil. Positively charged bases include calcium, magnesium, potassium, sodium, ammonium nitrogen, and several trace minerals.

When optimum ratios of bases exist, soil is believed to support high biological activity, have optimal physical properties (water intake and aggregation), and become resistant to leaching. Plants growing on such a soil are also balanced in mineral levels and are considered to be nutritious to humans and animals alike. Base saturation percentages that optimal for the growth of most crops are:

Calcium	60–70%
Magnesium	10–20%
Potassium	2–5%
Sodium	0.5–3%
Other bases	5%

According to Albrecht, fertilizer and lime applications should be made at rates that will bring soil mineral percentages into this ideal range. This approach will shift the soil pH automatically into a desirable range without creating nutrient imbalances. The base saturation theory also takes into account the effect one nutrient may have on another and avoids undesirable interactions.

For example, phosphorus is known to tie up zinc. The Albrecht system of soil evaluation contrasts with the approach used by many state laboratories, often called the sufficiency method. Sufficiency theory places little to no value on nutrient ratios, and lime recommendations are typically based on pH measurements alone. While in many circumstances base saturation and sufficiency methods will produce identical soil recommendations and similar results, significant differences can occur on a number of soils.

For example, suppose we tested a cornfield and found a soil pH of 5.5 and base saturation for magnesium at 20% and calcium at 40%. Base saturation theory would call for liming with a high-calcium lime to raise the per cent base saturation of calcium; the pH would rise accordingly. Sufficiency theory would not specify high-calcium lime and the grower might choose instead a high-magnesium dolomite lime that would raise the pH but worsen the balance of nutrients in soil.

Another way to look at these two theories is that the base saturation theory does not concern itself with pH to any great extent, but rather with the proportional amounts of bases. The pH will be correct when the levels of bases are correct. Albrecht's ideas have found their way onto large numbers of American farms and into the programmes of several agricultural consulting companies.

Conventional Fertilizers

Commercial fertilizer can be a valuable resource to farmers in transition to a more sustainable system and can help meet nutrient needs during times of high crop nutrient demand or when weather conditions result in slow nutrient release from organic resources. Commercial fertilizers have the advantage of supplying plants with immediately available forms of nutrients. They are often less expensive and less bulky to apply than many natural fertilizers. Not all conventional fertilizers are alike. Many appear harmless to soil livestock, but some are not. Anhydrous ammonia contains approximately 82% nitrogen and is applied subsurface as a gas. Anhydrous speeds the decomposition of organic matter in soil, leaving soil more compact as a result.

The addition of anhydrous causes increased acidity in soil, requiring 148 pounds of lime to neutralize 100 pounds of anhydrous ammonia, or 1.8 pounds of lime for every pound of nitrogen contained in the anhydrous.

Anhydrous ammonia initially kills many soil microorganisms in the application zone. Bacteria and actinomycetes recover within one to two weeks to levels higher than those prior to treatment. Soil fungi, however, may take seven weeks to recover. During the recovery time, bacteria are stimulated to grow more, and decompose more organic matter, by the high soil nitrogen content.

As a result, their numbers increase after anhydrous applications, then decline as available soil organic matter is depleted. Farmers commonly report that the long-term use of synthetic fertilizers, especially anhydrous ammonia, leads to soil compaction and poor tilth. When bacterial populations and soil organic matter decrease, aggregation declines, because existing glues that stick soil particles together are degraded, and no other glues are being produced. Potassium chloride (KCl) (0-0-60 and 0-0-50), also known as muriate of potash, contains approximately 50 to 60% potassium and 47.5% chloride.

Muriate of potash is made by refining potassium chloride ore, which is a mixture of potassium and sodium salts and clay from the brines of dying lakes and seas. The potential harmful effects from KCl can be surmised from the salt concentration of the material. Additionally, some plants such as tobacco, potatoes, peaches, and some legumes are especially sensitive to chloride. High rates of KCl must be avoided on such crops.

Potassium sulfate, potassium nitrate, sul-po-mag, or organic sources of potassium may be considered as alternatives to KCl for fertilization. Sodium nitrate, also known as Chilean nitrate or nitrate of soda, is another high-salt fertilizer. Because of the relatively low nitrogen content of sodium nitrate, a high amount of sodium is added to the soil when normal applications of nitrogen are made with this material. The concern is that excessive sodium acts as a dispersant of soil particles, degrading aggregation.

TOP SOIL – YOUR FARM'S CAPITAL

Topsoil is the capital reserve of every farm. Ever since mankind started agriculture, erosion of topsoil has been the single largest threat to a soil's productivity, and consequently, to farm profitability. This is still true today. In the US, the average acre of cropland is eroding at a rate of 7 tons per year. To sustain agriculture means to sustain soil resources, because that's the source of a farmer's livelihood.

The major productivity costs to the farm associated with soil erosion come from the replacement of lost nutrients and reduced water holding ability, accounting for 50 to 75% of productivity loss. Soil that is removed by erosion typically contains about three times more nutrients than the soil left behind and is 1.5 to 5 times richer in organic matter.

This organic matter loss not only results in reduced water-holding capacity and degraded soil aggregation, but also loss of plant nutrients, which must then be replaced with nutrient amendments. Five tons of topsoil (the so-called tolerance level) can easily contain 100 pounds of nitrogen, 60 pounds of phosphate, 45 pounds of potash, 2 pounds of calcium, 10 pounds of magnesium, and 8 pounds of sulfur.

The effect of slight, moderate, and severe erosion on organic matter, soil phosphorus level, and plantavailable water on a silt loam soil in Indiana. Water erosion gets started when falling rainwater collides with bare ground and detaches soil particles from the parent soil body. After enough water builds up on the soil surface, following detachment, overland water flow transports suspended soil down-slope. Suspended soil in the runoff water abrades and detaches additional soil particles as the water travels overland.

Table:7. Salt Index for Various Fertilizers

Material	*Salt Index*	*Salt Index per Unit of Plant Food*
Sodium Chloride	153	2.9
Patassium Chloride	116	1.9
Ammoniumn Nitrate	105	3.0
Sodium Nitrate	100	6.1
Urea	75	1.6
Potassium Nitrate	74	1.6
Ammonium Sulfate	69	3.3
Calcium Nitrate	53	4.4
Anhydrous Ammonia	47	0.6
Sulfate-Potash-Magnesia	43	2.0
Di-Ammonium Phosphate	34	1.6
Monammonium Phosphate	30	2.5
Gypsum	8	0.3
Calcium Carbonate	5	0.1

Preventing detachment is the most effective point of erosion control because it keeps soil in place. Other erosion control practices seek to slow soil particle transport and cause soil to be deposited before it reaches streams. These methods are less effective at protecting the quality of soil within the field. Commonly implemented practices to slow soil transport include terraces and diversions.

Terraces, diversions, and many other erosion, control, practices are largely unnecessary if the ground stays covered year-round. For erosion prevention, a high percentage of ground cover is a good indicator of success, while bare ground is an, early warning, indicator for a high risk of erosion. Muddy runoff water and gullies are too-late, indicators. Soil has already eroded by the time it shows up as muddy water, and it's too late to save soil already suspended in the water.

Protecting soil from erosion is *the* first step toward a sustainable agriculture. Since water erosion is initiated by raindrop impact on bare soil, any management practice that protects soil from raindrop impact will decrease erosion and increase water entry into soil. Mulches, cover crops, and crop residues serve this purpose well.

Fig:9. Raindrops Falling on Bare Ground Initiate Erosion
Drawing from Cropland Monitoring Quide

Additionally, well-aggregated soils resist crusting because water-stable aggregates are less likely to break apart when the raindrop hits them. Adequate organic matter with high soil biological activity leads to high soil aggregation. Many studies have shown that cropping systems that maintain a soil-protecting plant canopy or residue cover have the least soil erosion. This is universally true.

Long-term cropping studies begun in 1888 at the University of Missouri provide dramatic evidence of this. Gantzer and colleagues examined the effects of a century of cropping on soil erosion. They compared depth of topsoil remaining after 100 years of cropping. The cropping system that maintained the highest amount of permanent ground cover (timothy grass) had the greatest amount of topsoil left.

The researchers commented that subsoil had been mixed with topsoil in the continuous corn plots from plowing, making the real topsoil depth less than was apparent. In reality, all the topsoil was lost from the continuous corn plots in only 100 years. The rotation lost about half the topsoil over 100 years. How can we feed future generations with this type of farming practice? In a study of many different soil types in each of the major climatic zones of the US, researchers showed dramatic differences in soil erosion when comparing row crops to perennial sods. Row crops consisted of cotton or corn, and sod crops were bluegrass or bermuda grass. On average, the row crops eroded more than 50 times more soil than did the perennial sod crops. The two primary influencing factors are ground cover and tillage. So, how long do fields have before the topsoil is gone? This depends on where in the country the field is located. Some soils naturally have very thick topsoil, while other soils have thin topsoil over rock or gravel. Roughly 8 tons/acre/year of soil-erosion loss amounts to the thickness of a dime spread over an acre.

Twenty dimes stack up to 1-inch high. So a landscape with an 8-ton erosion rate would lose an inch of topsoil about every 20 years. On a soil with a thick topsoil, this amount is barely detectable within a person's lifetime and may not be no ticed. Soils with naturally thin topsoils or topsoils that have been previously eroded can be transformed from productive to degraded land within a generation. Forward-thinking researcher Wes Jackson, of the Land Institute, waxes eloquent about how tillage has become engrained in human culture since we first began farming. Beating our swords into plowshares surely embodies the triumph of good over evil.

Someone who creates something new is said to have plowed new ground. Yet the plowshare may well have destroyed more options for future generations than the sword. Tillage for the production of annual crops is the major problem in agriculture, causing soil erosion and the loss of soil quality. Any agricultural practice that creates and maintains bare ground is inherently less sustainable than practices that keep the ground covered throughout the year.

Wes Jackson has spent much of his career developing perennial grain crops and cropping systems that mimic the natural Prairie. Perennial grain crops do not require tillage to establish year after year, and the ground is left covered. Ultimately, this is the future of grain production and truly represents a new vision for how we produce food. The greatest research need in agriculture today is breeding work to develop perennial crops that will replace annual crops requiring tillage. Farming practices using annual crops in ways that mimic perennial systems, such as no-till and cover crops, are our best alternative until perennial systems are developed.

Chapter 3

Mycorrhiza in Soil

INTRODUCTION

Mycorrhiza refers to an association or symbiosis between plants and fungi that colonise the cortical tissue of roots during periods of active plant growth. The association is characterised by the movement of plant-produced carbon to the fungus and fungal-acquired nutrients to the plant. The term mycorrhiza, which literally means *fungus-root*, was first applied to fungus-tree associations described in 1885 by the German forest pathologist A.B. Frank. Since then we have learned that the vast majority of land plants form symbiotic associations with fungi: an estimated 95% of all plant species belong to genera that characteristically form mycorrhizae. The mycorrhizal condition is the rule among plants, not the exception.

The benefits afforded plants from mycorrhizal symbioses can be characterised either agronomically by increased growth and yield or ecologically by improved fitness (i.e., reproductive ability). In either case, the benefit accrues primarily because mycorrhizal fungi form a critical linkage between plant roots and soil. Mycorrhizal fungi usually proliferate both in the root and in soil. The soilborne or extramatrical hyphae take up nutrients from soil solution and transport them to the root. By this mechanism, mycorrhizae increase the effective absorptive surface area of the plant.

In nutrient-poor or moisture-deficient soils, nutrients taken up by the extramatrical hyphae can lead to improved plant growth and reproduction. As a result, mycorrhizal plants are often more competitive and better able to tolerate environmental stresses than are non-mycorrhizal plants.

GLOBAL PERSPECTIVE

Mycorrhizal associations vary widely in structure and function. Despite the many exceptions, it is possible to state broad generalisations about latitude (or altitude), soil properties, and structure and function of the different mycorrhizal types that colonize the dominant vegetation in a gradient of climatic zones. Ericaceous plants, which dominate the acidic, high-organic heathland soils of subarctic and subalpine regions, are colonised by a group of ascomycetous fungi, giving rise to the ericoid type of mycorrhiza. This mycorrhizal type is characterised by extensive growth within (i.e., intracellular) cortical cells, but little extension into soil. The fungi produce extracellular enzymes that break down organic matter, enabling the plant to assimilate nutrients mineralised from organic compounds present in the colloidal material surrounding roots. Moving along the environmental gradient, coniferous trees replace ericaceous shrubs as the dominant vegetation. These trees are colonised by a wide range of mostly basidiomycetous fungi that grow between (i.e., intercellular) root cortical cells forming the ectomycorrhizal type of mycorrhiza.

Ectomycorrhizal fungi may produce large quantities of hyphae on the root and in soil. These hyphae function in the absorption and translocation of inorganic nutrients and water, but also release nutrients from litter layers by production of enzymes involved in mineralisation of organic matter. At the warmer and drier end of the environmental gradient, grasslands often form the dominant vegetation. In these ecosystems nutrient use is high and phosphorus is frequently a limiting element for growth. Grasses and a wide variety of other plants are colonised by fungi belonging to the order Glomales. These fungi form arbuscules or highly branched structures within (intracellular) root cortical cells, giving rise to the arbuscular type of mycorrhiza.

The Glomalean fungi may produce extensive extramatrical hyphae (i.e., hyphae outside the root) and can significantly increase phosphorus-inflow rates of the plants they colonise. The diversity of these root-fungal associations provides plants with a range of strategies for efficient functioning in an array of plant-soil systems. The objective of this paper is to provide an overview of this diversity and to evaluate the roles and potential for management, of the mycorrhizal symbioses in native and managed ecosystems. Because ectomycorrhizae and arbuscular mycorrhizae are the most widespread, we will emphasise these types of associations.

TYPES OF MYCORRHIZAE

Ectomycorrhizae

The diagnostic feature of ectomycorrhizae (EM) is the presence of hyphae between root cortical cells producing a netlike structure called the Hartig net, after Robert Hartig who is considered the father of forest biology. Many EM also have a sheath, or mantle, of fungal tissue that may completely cover the absorbing root (usually the fine feeder roots). The mantle can vary widely in thickness, colour, and texture depending on the particular plant-fungus combination. The mantle increases the surface area of absorbing roots and often affects fine-root morphology, resulting in root bifurcation and clustering. Contiguous with the mantle are hyphal strands that extend into soil. Often the hyphal strands will aggregate to form rhizomorphs that may be visible to the unaided eye. The internal portion of rhizomorphs can differentiate into tubelike structures specialised for long-distance transport of nutrients and water.

Ectomycorrhizae are found on woody plants ranging from shrubs to forest trees. Many of the host plants belong to the families Pinaceae, Fagaceae, Betulaceae and Myrtaceae. Over 4,000 fungal species, belonging primarily to the Basidiomycotina, and fewer to the Ascomycotina, are known to form ectomycorrhizae. Many of these fungi produce mushrooms and puffballs on the forest floor. Some fungi have a narrow host range, such as *Boletus betulicola* on *Betula* spp., while others have very broad host range, such as *Pisolithus arhizus* (also called *P. tinctorius*) which forms ectomycorrhiza with more than 46 tree species belonging to at least eight genera.

Arbuscular Mycorrhizae

The diagnostic feature of arbuscular mycorrhizae (AM) is the development of a highly branched arbuscule within root cortical cells. The fungus initially grows between cortical cells, but soon penetrates the host cell wall and grows within the cell. The general term for all mycorrhizal types where the fungus grows within cortical cells is endomycorrhiza. In this association neither the fungal cell wall nor the host cell membrane are breached.

As the fungus grows, the host cell membrane invaginates and envelops the fungus, creating a new compartment where material of high molecular complexity is deposited. This apoplastic space prevents direct contact between the plant and fungus cytoplasm and allows for efficient transfer of nutrients between the symbionts. The arbuscules are relatively short lived, less than 15 days, and are often difficult to see in field-collected samples.

Other structures produced by some AM fungi include vesicles, auxiliary cells, and asexual spores. Vesicles are thin-walled, lipid-filled structures that usually form in intercellular spaces. Their primary function is thought to be for storage; however, vesicles can also serve as reproductive propagules for the fungus. Auxiliary cells are formed in the soil and can be coiled or knobby. The function of these structures is unknown. Reproductive spores can be formed either in the root or more commonly in the soil. Spores produced by fungi forming AM associations are asexual, forming by the differentiation of vegetative hyphae. For some fungi (e.g., *Glomus intraradices*), vesicles in the root undergo secondary thickening, and a septum (cross wall) is laid down across the hyphal attachment leading to spore formation, but more often spores develop in soil from hyphal swellings.

The fungi that form AM are currently all classified in the order Glomales. The taxonomy is further divided into suborders based on the presence of:

- Vesicles in the root and formation of chlamydospores (thick wall, asexual spore) borne from subtending hyphae for the suborder Glomineae or
- Absence of vesicles in the root and formation of auxiliary cells and azygospores (spores resembling a zygospore but developing asexually from a subtending hypha resulting in a distinct bulbous attachment) in the soil for the suborder Gigasporineae.

The term vesicular-arbuscular mycorrhiza (VAM) was originally applied to symbiotic associations formed by all fungi in the Glomales, but because a major suborder lacks the ability to form vesicles in roots, AM is now the preferred acronym. The order Glomales is further divided into families and genera according to the method of spore formation. The spores of AM fungi are very distinctive.

They range in diameter from 10 mm. for *Glomus tenue* to more than 1,000 mm. for some *Scutellospora* spp. The spores can vary in colour from hyaline (clear) to black and in surface texture from smooth to highly ornamented. *Glomus* forms spores on the ends of hyphae, *Acaulospora* forms spores laterally from the neck of a swollen hyphal terminus, and *Entrophospora* forms spores within the neck of the hyphal terminus.

The Gigasporineae are divided into two genera based upon the presence of inner membranous walls and a germination shield (wall structure from which the germ tube can arise) for *Scutellospora* or the absence of these structures for *Gigaspora*. The AM type of symbiosis is very common as the fungi involved can colonise a vast taxonomic range of both herbaceous and woody plants, indicating a general lack of host specificity among this type. However, it is important to distinguish between *specificity*, innate ability to colonise, infectiveness, amount of colonization, and effectiveness, plant response to colonisation. AM fungi differ widely in the level of colonisation they produce in a root system and in their impact on nutrient uptake and plant growth.

Ericaceous Mycorrhizae

The term ericaeous is applied to mycorrhizal associations found on plants in the order Ericales. The hyphae in the root can penetrate cortical cells (endomycorrhizal habit); however, no arbuscules are formed. Three major forms of ericaceous mycorrhiza have been described:

- *Ericoid*: Cells of the inner cortex become packed with fungal hyphae. A loose welt of hyphae grows over the root surface, but a true mantle is not formed. The ericoid mycorrhizae are found on plants such as *Calluna* (heather), *Rhododendron* (azaleas and rhododendrons) and *Vaccinium* (blueberries) that have very fine root systems and typically grow in acid, peaty soils. The fungi involved are ascomycetes of the genus *Hymenoscyphus*.
- *Arbutoid:* Characteristics of both EM and endomycorrhizae are found. Intracellular penetration can occur, a mantle forms, and a Hartig net is present. These associations are found on *Arbutus* (e.g., Pacific madrone), *Arctostaphylos* (e.g., bearberry), and several species of the Pyrolaceae. The fungi involved in the association are basidiomycetes and may be the same fungi that colonise EM tree hosts in the same region.
- *Monotropoid*: The fungi colonise achlorophyllous (lacking chlorophyll) plants in Monotropaceae (e.g., Indian pipe), producing the Hartig net and mantle. The same fungi also form EM associations with trees and thereby form a link through which carbon and other nutrients can flow from the autotrophic host plant to the heterotrophic, parasitic plant.

Orchidaceous Mycorrhizae

Mycorrhizal fungi have a unique role in the life cycle of plants in the Orchidaceae. Orchids typically have very small seeds with little nutrient reserve. The plant becomes colonized shortly after germination, and the mycorrhizal fungus supplies carbon and vitamins to the developing embryo. For achlorophyllous species, the plant depends on the fungal partner to supply carbon throughout its life.

The fungus grows into the plant cell, invaginating the cell membrane and forming hyphal coils within the cell. These coils are active for only a few days, after which they lose turgor and degenerate and the nutrient contents are absorbed by the developing orchid. The fungi participating in the symbiosis are basidiomycetes similar to those involved in decaying wood (e.g., *Coriolus, Fomes, Marasmius*) and pathogenesis (e.g., *Armillaria* and *Rhizoctonia*). In mature orchids, mycorrhizae also have roles in nutrient uptake and translocation.

Mixed Infections

Several fungi can colonise the roots of a single plant, but the type of mycorrhiza formed is usually uniform for a host. In some cases, however, a host can support more than one type of mycorrhizal association. *Alnus* (alders), *Salix* (willows), *Populus* (poplars), and *Eucalyuptus* can have both AM and EM associations on the same plant. Some ericoid plants have occasional EM and AM colonization. An intermediate mycorrhizal type can be found on coniferous and deciduous hosts in nurseries and burned forest sites. The ectendomycorrhiza type forms a typical EM structure, except the mantle is thin or lacking and hyphae in the Hartig net may penetrate root cortical cells. The fungi involved in the association were initially designated "E-strain".

UPTAKE AND TRANSFER OF SOIL NUTRIENTS

When a nutrient is deficient in soil solution, the critical root parameter controlling its uptake is surface area. Hyphae of mycorrhizal fungi have the potential to greatly increase the absorbing surface area of the root while extramatrical mycelia (aggregates of hyphae) accounted for less than 20% of the total nutrient absorbing surface mass, they contributed nearly 80% of the absorbing surface area of pine seedlings.

It is also important to consider the distribution and function of the extramatrical hyphae. If the mycorrhiza is to be effective in nutrient uptake, the hyphae must be distributed beyond the nutrient depletion zone that develops around the root. A nutrient depletion zone develops when nutrients are removed from the soil solution more rapidly than they can be replaced by diffusion. For a poorly-mobile ion such as phosphate, a sharp and narrow depletion zone develops close to the root. Hyphae can readily bridge this depletion zone and grow into soil with an adequate supply of phosphorus. Uptake of micronutrients such as zinc and copper is also improved by mycorrhizae because these elements are also diffusion-limited in many soils. For more mobile nutrients such as nitrate, the depletion zone is wide and it is less likely that hyphae grow extensively into the zone that is not influenced by the root alone.

Another factor contributing to the effective absorption of nutrients by mycorrhizae is their narrow diameter relative to roots. The steepness of the diffusion gradient for a nutrient is inversely related to the radius of the absorbing unit; therefore, the soil solution should be less depleted at the surface of a narrow absorbing unit such as a hypha. Furthermore, narrow hyphae can grow into small soil pores inaccessible to roots or even root hairs. Another advantage attributed to mycorrhizal fungi is access to pools of phosphorus not readily available to the plant. One mechanism for this access is the physiochemical release of inorganic and organic phosphorus by organic acids through the action of low-molecular-weight organic anions such as oxalate.

- Replace phosphorus sorbed at metal-hydroxide surfaces through ligand-exchange reactions.
- Dissolve metal-oxide surfaces that sorb phosphorus.
- Complex metals in solution and thus prevent precipitation of metal phosphates.

Some EM fungi produce large quantities of oxalic acid, and this may partially explain enhanced nutrient uptake by EM roots. Another mechanism by which mycorrhizal fungi release inorganic phosphorus is through mineralisation of organic matter. This occurs by phosphatase-mediated hydrolysis of organic phosphate (C-O-P) ester bonds.

Significant phosphatase activity has been documented for mycorrhizal fungi grown in pure cultures and for excised and intact EM short roots. In the field, a positive correlation has been reported between phosphatase activity and the length of fungal hyphae associated with EM mantles. Care must be exercised in interpreting these data because plant roots and the associated microflora also produce organic acids and phosphatases; however, mycorrhizal fungi certainly intensify this activity.

Ericoid and EM have a special role in the mineralisation of nitrogen. Most plant litter entering the soil has a high C:N ratio and is rich in lignin and tannins. Only a few mycorrhizal fungi can mobilise nutrients from these primary sources. However, a wide range of ericoid and EM fungi can obtain nitrogen and other nutrients from secondary sources of organic matter such as dead microbial biomass. A wide range of hydrolytic and oxidative enzymes capable of depolymerising organic nitrogen have been demonstrated. These types of mycorrhizae may have an important role in nitrogen cycling in the acidic and highly organic soils where they predominate.

CARBON FLUXES IN MYCORRHIZAL PLANTS

Mycorrhizal fungi range from obligate symbionts, which can only obtain carbon from the plant host as in the case of AM fungi to facultative symbionts, which can also mineralise organic carbon from nonliving sources as in the case of some EM species. In nature the heterotrophic mycorrhizal fungi obtain all or most of their carbon from the autotrophic host plant. Ectomycorrhizae and ericoid mycorrhizae transform host carbohydrates into fungal-specific storage carbohydrates, such as mannitol and trehalose, which may produce a sink for photosynthate that favours transport of carbohydrate to the fungal partner. In AM, lipids accumulate in vesicles and other fungal structures and provide an analogous sink for host photosynthate.

As much as 20% of the total carbon assimilated by plants may be transferred to the fungal partner. This transfer of carbon to the fungus has sometimes been considered a drain on the host. However, the host plant may increase photosynthetic activity following mycorrhizal colonisation, thereby compensating for carbon "lost" to the soil. Occasionally, plant growth suppression has been attributed to mycorrhizal colonization, but usually this occurs only under low-light (photosynthate limiting) or high-phosphorus conditions.

In an ecosystem, the flow of carbon to the soil mediated by mycorrhizae serves several important functions. For some mycorrhizae, the extramatrical hyphae produce hydrolytic enzymes, such as proteases and phosphatases, that can have an important impact on organic matter mineralisation and nutrient availability. Extramatrical hyphae of mycorrhizae also bind soil particles together and thereby improve soil aggregation. Typically there are between 1 to 20 m of AM hyphae gof soil. Another important consequence of carbon flow to the fungal partner is the development of a unique rhizosphere microbial community called the mycorrhizosphere shortly. Soil scientists now realise that carbon flow to soil is critical for the development of soil aggregation and the maintenance of a healthy plant-soil system. Enhanced carbon flow to the soil should be considered an important benefit of mycorrhizal colonisation.

INTERACTIONS WITH OTHER SOIL ORGANISMS

Mycorrhizal fungi interact with a wide assortment of organisms in the rhizosphere. The result can be either positive, neutral, or negative on the mycorrhizal association or a particular component of the rhizosphere. For example, specific bacteria stimulate EM formation in conifer nurseries and are called mycorrhization helper bacteria. In certain cases these bacteria eliminate the need for soil fumigation.

The interaction between rhizobia and AM fungi has received considerable attention because of the relatively high phosphorus demand of N_2 fixation. The two symbioses typically act synergistically, resulting in greater nitrogen and phosphorus content in combination than when each is inoculated onto the legume alone. Legumes are typically coarse-rooted and therefore inefficient in extracting phosphorus from soil. The AM fungi associated with legumes are an essential link for adequate phosphorus nutrition, leading to enhanced nitrogenase activity that in turn promotes root and mycorrhizal growth.

Mycorrhizal fungi colonize feeder roots and thereby interact with root pathogens that parasitise this same tissue. In a natural ecosystem where the uptake of phosphorus is low, a major role of mycorrhizal fungi may be protection of the root system from endemic pathogens such as *Fusarium* spp. Mycorrhizae may stimulate root colonisation by selected biocontrol agents, but our understanding of these interactions is meager. Much more research has been conducted on the potential effects of mycorrhizal colonisation on root pathogens. Mycorrhizal fungi may reduce the incidence and severity of root diseases. The mechanisms proposed to explain this protective effect include:

- Development of a mechanical barrier-especially the mantle of the EM-to infection by pathogens.
- Production of antibiotic compounds that suppress the pathogen.

- Competition for nutrients with the pathogen, including production of siderophores.
- Induction of generalised host defence mechanisms.

MANAGEMENT OF MYCORRHIZAE

The dramatic plant growth response achieved in pot studies following inoculation with mycorrhizal fungi in low-fertility soils led to a flurry of activity in the 1980s aimed at using these organisms as biofertilizers. Field responses were often disappointing, especially in high-input agricultural systems, and many concluded that mycorrhizae had little practical importance in agriculture. Further studies, however, have confirmed that most agricultural plants are colonised by mycorrhizal fungi and that they can have a substantial impact, both positive and negative, on crop productivity.

Certainly, agriculturists should appreciate the distribution of mycorrhizae within their systems and understand the impact of their management decisions on mycorrhizal functioning. Factors that should be considered when assessing the potential role of mycorrhizae in an agroecosystem include:

Mycorrhizal Dependency (MD) of the Host Crop

This is usually defined as the growth response of mycorrhizal (M) versus non-mycorrhizal (NM) plants at a given phosphorus level; MD = ((M – NM)/NM) × 100. Although most agricultural crops have mycorrhizae, not all benefit equally from the symbiosis. Generally, coarse-rooted plants benefit more than fine-rooted plants.

Nutrient Status of Soil

Assuming that the major benefit of the mycorrhizal symbiosis is improved phosphorus uptake, the management of mycorrhizal fungi will be most critical when soil phosphorus is limiting. Many tropical soils fix phosphorus and proper mycorrhization of plants is essential to obtain adequate phosphorus nutrition. In temperate zones, phosphorus is sometimes applied in excess of crop demand. However, with increased concerns about environmental quality, phosphorus use in developed countries may be reduced, resulting in increased dependence on native mycorrhizae for nutrient uptake. Another factor to consider is the interaction of water stress with nutrient availability. As soils dry, phosphorus may become limiting even in soils that test high in available phosphorus.

Inoculum Potential of the Indigenous Mycorrhizal Fungi

Inoculum potential is a product of the *abundance* and *vigour* of the propagules in the soil and can be quantified by determining the rate of colonisation of a susceptible host under a standard set of conditions. Inoculum potential can be adversely affected by management practices such as fertilizer and lime application, pesticide use, crop rotation, fallowing, tillage, and topsoil removal.

Mycorrhizal Populations in Soil

Soil disturbance such as tillage can dramatically affect the function of mycorrhizae in an agricultural system. M.H. Miller and coworkers from the University of Guelph, Canada, documented an interesting case where disturbance of an arable, no-till soil resulted in reduced AM development and subsequently less absorption of phosphorus by seedlings of maize in the field. They hypothesised that soil disturbance reduced the effectiveness of the mycorrhizal symbiosis. To confirm this, they conducted a series of growth chamber studies with nondisturbed and disturbed soil cores collected from long-term field plots. Disturbance reduced both mycorrhizal colonization and phosphorus absorption by maize and wheat roots, but did not reduce phosphorus absorption by two non-mycorrhizal crops, spinach and canola. The authors concluded that under nutrient-limited conditions, the ability of mycorrhizal seedlings to associate with intact hyphal networks in soil may be highly advantageous for crop establishment.

Crop rotation and fallow systems can affect the diversity and function of mycorrhizal fungi. In semiarid cropping systems, clean fallows conserve soil moisture and nitrate for the subsequent crop. Since the 1940s, some crops sown immediately after long fallow grew poorly and had phosphorus and zinc deficiencies. They conducted inoculation trials and found that increasing inoculum abundance in soil overcame the deleterious impact of fallow. They recommended that farmers avoid planting mycorrhizal-dependent crops, such as linseed, sunflower, and soybean, following periods of fallow or after a nonhost plant such as canola that lead to reduction in AM propagules.

POTENTIAL FOR INOCULUM PRODUCTION AND USE

In situations where native mycorrhizal inoculum potential is low or ineffective, providing the appropriate fungi for the plant production system is worth considering. With the current state of technology, inoculation is best for transplanted crops and in areas where soil disturbance has reduced native inoculum potential.

The first step in any inoculation programme is to obtain an isolate that is both infective, or able to penetrate and spread in the root, and effective, or able to enhance growth or stress tolerance of the host. Individual isolates of mycorrhizal fungi vary widely in these properties, so screening trials are important to select isolates that will perform successfully. Screening under actual cropping conditions is best because indigenous mycorrhizal fungi, pathogens, and soil chemical and physical properties will influence the result.

Isolation and inoculum production of EM and AM fungi present very different problems. Many EM fungi can be cultured on artificial media. Therefore, isolates of EM fungi can be obtained by placing surface-disinfested portions of sporocarps or mycorrhizal short roots on an agar growth medium. The resulting fungal biomass can be used directly as inoculum but, for ease of use, inoculum often consists of the fungal material mixed with a carrier or bulking material such as peat. Obtaining isolates of AM fungi is more difficult because they will not grow apart from their hosts.

Spores can be sieved from soil, surface disinfested, and used to initiate "pot cultures" on a susceptible host plant in sterile soil or an artificial plant-growth medium. Inoculum is typically produced in scaled-up pot cultures. Alternatively, hydroponic or aeroponic culture systems are possible; a benefit of these systems is that plants can be grown without a supporting substratum, allowing colonised roots to be sheared into an inoculum of high propagule number. Sylvia summarised methods for working with AM inoculum examples where inoculating with either EM or AM fungi is beneficial when planting a mycorrhizal-dependent crop in an area where native inoculum potential is low:

Pines were not native to Puerto Rico, and their fungal symbionts were absent from the soil. As far back as the 1930s attempts to establish pine on the island were unsuccessful. Typically, the pines germinated well and grew to heights of 8 to 10 cm. in a relatively short time, but then rapidly declined. Phosphorus fertilizers did not substantially improve plant vigour. In 1955, soil from under pine stands in North Carolina was transported to Puerto Rico where it was incorporated as inoculum into soil around 1-year-old "scrawny" pine seedlings growing at Maricao in the western mountains.

Thirty-two seedlings were inoculated, and an equal number were monitored as noninoculated control plants. Within one year, inoculated plants had abundant mycorrhizal colonisation and had achieved heights of up to 1.5 m., while most of the noninoculated plants had died. Further trials with mixtures of surface soil containing mycorrhizal fungi and with pure inocula, consisting of fungi growing in a peat-based medium, confirmed that inoculated seedlings were consistently more vigorous and larger than nonmycorrhizal ones. Subsequent surveys more than 15 years after inoculation indicated that the inoculated fungi continued to grow and sporulate in the pine plantations.

Beach erosion is a problem in many coastal areas and replenishing the beaches with sand dredged from offshore is often the method of choice for restoring them. Native grasses are planted in the back beach to reduce further erosion and to initiate the dune-building process. In native dunes, beach grasses are colonised by a wide array of AM fungi. However, when these grasses are propagated in nurseries, they do not have mycorrhizae. Furthermore, the replenishment sand is typically devoid of AM propagules. AM fungi were isolated from grasses growing in native dunes.

The fungi were screened for effectiveness with the given host/soil combination and for compatibility with the nursery production system, and the effect of inoculation was documented on transplants placed on newly restored beaches. In the nursery, moderate amounts of colonisation were achieved, even with high levels of pesticide and fertilizer use. After transfer of these plants to a low-nutrient beach environment, AM colonisation spread rapidly and enhanced plant growth significantly compared to noninoculated control plants even though plants were equal size when they left the intensively managed nursery.

Compared to non-inoculated plants after 20 months on the beach, AM-colonised plants had 219, 81, 64 and 53% more shoot dry mass, root length, plant height, and number of tillers, respectively. In most cases the objective of nursery inoculation is not to achieve a growth response, but rather to establish the symbiosis with the plant so that it can be effectively transferred to the field.

ARBUSCULAR MYCORRHIZA

An arbuscular mycorrhiza (plural mycorrhizae or mycorrhizas) is a type of mycorrhiza in which the fungus penetrates the cortical cells of the roots of a vascular plant. Arbuscular mycorrhizae (AMs) are characterised by the formation of unique structures such as arbuscules and vesicles by fungi of the phylum Glomeromycota (AM fungi). AM fungi help plants to capture nutrients such as phosphorus and micronutrients from soil. It is believed that the development of the arbuscular mycorrhizal symbiosis played a crucial role in the initial colonisation of land by plants and in the evolution of the vascular plants.

It has been said that it is quicker to list the plants that do not form mycorrhizae than those that do. This symbiosis is a highly evolved mutualistic relationship found between fungi and plants, the most prevalent plant symbiosis known, and AM is found in of 80% of vascular plant families of today. The tremendous advances in research on mycorrhizal physiology and ecology over the past 40 years have led to a greater understanding of the multiple roles of AMF in the ecosystem. This knowledge is applicable to human endeavours of ecosystem management, ecosystem restoration and agriculture.

EVOLUTION OF MYCORRHIZAL SYMBIOSIS

Paleobiology

Both paleobiological and molecular evidence indicate that AM is an ancient symbiosis that originated at least 460 million years ago. AM symbiosis is ubiquitous among land plants, which suggests that mycorrhizae were present in the early ancestors of extant land plants. This positive association with plants may have facilitated the development of land plants.

The Rhynie Chert of the lower Devonian has yielded fossils of the earliest land plants in which AM fungi have been observed. The fossilised plants containing mycorrhizal fungi were preserved in silica.

The Early Devonian saw the development of terrestrial flora. Plants of the Rhynie chert from the Lower Devonian (400 m.yrs. ago) were found to contain structures resembling vesicles and spores of present *Glomus* species. Colonised fossil roots have been observed in *Aglaophyton major* and *Rhynia*, which are ancient plants possessing characteristics of vascular plants and bryophytes with primitive protostelic rhizomes.

Intraradical mycelium was observed in root intracellular spaces, and arbuscules were observed in the layer thin wall cells similar to palisade parenchyma. The fossil arbuscules appear very similar to those of existing AMF. The cells containing arbuscules have thickened walls which are also observed in extant colonised cells. Mycorrhizae from the Miocene exhibit a vesicular morphology closely resembling that of present *Glomerales*. The need for further evolution may have been lost due to the readily available food source provided by the plant host. However, it can be argued that the efficacy of signaling process is likely to have evolved and this could be not be easily detected in the fossil record. A fine tuning of the signaling processes would improve coordination and nutrient exchange between symbionts and increasing the fitness of both the fungi and the plant symbionts.

The nature of the relationship between plants and the ancestors of arbuscular mycorrhizal fungi is contentious. Two hypotheses are that:

- Mycorrhizal symbiosis evolved from a parasitic interaction which developed in to a mutually beneficial relationship.
- Mycorrhizal fungi developed from saprobic fungi that became endosymbiotic.

Both saprotrophs and biotrophs were found in the Rhynie Chert but there is little evidence to support either hypothesis.

There is some fossil evidence that the parasitic fungi did not kill the host cells immediately upon invasion although a response to the invasion was observed in the host cells. This response may have evolved into the chemical signaling processes required for symbiosis. In both cases the symbiotic plant-fungi interaction is thought to have evolved from a relationship in which the fungi was taking nutrients from the plant into a symbiotic relationship where the plant and fungi exchange nutrients.

PHYSIOLOGY

Presymbiosis

The development of AM fungi prior to root colonisation, known as presymbiosis, consists of three stages: spore germination, hyphal growth, host recognition and appressorium formation. Spores of the AM fungi are thick walled multi-nucleate resting structures. The germination of the spore does not depend on the plant as spores have been germinated under experimental conditions in the absence of plants both *in vitro* and in soil. However, the rate of germination can be increased by host root exudates. AM fungal spores germinate given suitable conditions of the soil matrix, temperature, carbon dioxide concentration, pH and phosphorus concentration.The growth of AM hyphae through the soil is controlled by host root exudates and soil phosphorus concentration.

Low phosphorus concentrations in soil increase hyphal growth and branching as well as induce plant exudation of compounds which control hyphal branching intensity.The branching of AM fungal hyphae grown in 1 mM phosphorus media is significantly reduced but the length of the germ tube and total hyphal growth was not affected.

A concentration of 10 mM phosphorus inhibited both hyphal growth and branching. This phosphorus concentration occurs in natural soil conditions and could thus contribute to reduced mycorrhizal colonisation. Root exudates from AMF host plants grown in a liquid medium with and without phosphorus have been shown to effect hyphal growth. Pre-germinated surface-sterilised spores of *Gigaspora magarita* were grown in host plant exudates.

The fungi grow in the exudates from roots starved of phosphorus had increased hyphal growth and produced tertiary branches compared to those grown in exudates from plants given adequate phosphorus. When the growth promoting root exudates were added in low concentration the AM fungi produced scattered long branches. As the concentration of exudates was increased the fungi produced more tightly clustered branches. At the highest concentration arbuscules, the AMF structures of phosphorus exchange were formed.

This chemotaxic fungal response to the host plants exudates is thought to increase the efficacy of host root colonisation in low phosphorus soils. It is an adaptation for fungi to efficiently explore the soil in search of a suitable plant host. Further evidence that AM fungi exhibit host specific chemotaxis: spores of *Glomus mosseae* where separated from the roots of a host plant, non-host plants and dead host plant by a membrane only permeable to hyphae. In the treatment with the host plant the fungi crossed the membrane and always emerged within 800 μm of the root. Whereas in the treatments with non host plants and dead plants, the hyphae did not cross the membrane to reach the roots. This demonstrates that arbuscular mycorrhizal fungi have chemotaxic abilities which enable hyphal growth toward the roots of a potential host plant.

Molecular techniques have been used to further understand the signaling pathways which occur between arbuscular mycorrhizae and the plant roots. In the presence of exudates from potential host plant roots the AM undergoes physiological changes which allow it to colonise its host. AM fungal genes required for the respiration of spore carbon compounds are triggered and turned on by host plant root exudates. In experiments there was an increase in the transcription rate of 10 genes 0.5h after exposure and an even greater rate after 1h. A morphological growth response was observed 4 hours after exposure. The genes were isolated and found to be involved in mitochondrial activity and enzyme production.

The fungal respiration rate was measured by O_2 consumption rate and increased by 30% 3 hours after exposure to root exudates. This indicates that AMF spore mitochondrial activity is positively stimulated by host plant root exudates. This may be part of a fungal regulatory mechanism that conserves spore energy for efficient growth and the hyphal branching upon receiving signals from a potential host plant.

When arbuscular mycorrhizal fungal hyphae encounter the root of a host plant an apressorium (an infection structure) is formed on the root epidermis. The apressorium is the structure from which the hyphae can penetrate into the host's parenchyma cortex. The formation of apressoria does not require chemical signals from the plant. AM fungi could form apressoria on the cell walls of "ghost" cells in which the protoplast had been removed to eliminate signaling between the fungi and the plant host. However, the hyphae did not further penetrate the cells and grow in toward the root cortex which indicates that signaling between symbionts is required for further growth once appresoria are formed.

Symbiosis

Once inside the parenchyma the fungi forms highly branched structures for nutrient exchange with the plant called "arbuscules". These are the distinguishing structures of arbuscular mycorrhizal fungus. Arbuscules are the sites of exchange for phosphorus, carbon, water and other nutrients. The host plant exerts a control over the intercellular hyphal proliferation and arbuscule formation. There is a decondensation of the plant's chromatin which indicates increased transcription of the plant's

DNA in arbuscule containing cells. Major modifications are required in the plant host cell to accommodate the arbuscules. The vacuoles shrink and other cellular organelles proliferate. The plant cell cytoskeleton is reorganised around the arbuscules.

There are two other types of hyphae which originate from the colonised host plant root. Once colonisation has occurred short lived runner hyphae grow from the plant root into the soil. These are the hyphae that take up phosphorus and micronutrients which are conferred to the plant. AM fungal hyphae have a high surface to volume ratio making their absorptive ability greater than that of plant roots. AMF hyphae are also finer than roots and can enter into pores of soil that are inaccessible to roots. The third type of AMF hyphae grows from the roots and colonises other host plant roots. The three types of hyphae are morphologically distinct.

NUTRIENT UPTAKE AND EXCHANGE

AMF are obligate symbionts. They have limited saprobic ability and are dependent on the plant for their carbon nutrition. AM fungi take up the products of the plant host's photosynthesis as hexoses. The transfer of carbon from the plant to the fungi may occur through the arbuscules or intraradical hyphae. Secondary synthesis from the hexoses by AM occurs in the intraradical mycelium. Inside the mycelium, hexose is converted to trehalose and glycogen. Trehalose and glycogen are carbon storage forms which can be rapidly synthesized and degraded and may buffer the intracellular sugar concentrations. The intraradical hexose enters the oxidative pentose phosphate pathway which produces pentose for nucleic acids.

Lipid biosynthesis also occurs in the intraradical mycelium. Lipids are then stored or exported to extraradical hyphae where they may be stored or metabolised. The breakdown of lipids into hexoses, known as gluconeogenesis, occurs in the extraradical mycelium. Approximately 25% of the carbon translocated from the plant to the fungi is stored in the extraradical hyphae. Up to 20% of the host plant's photosynthate carbon may be transferred to the AM fungi. This represents a considerable carbon investment in mycorrhizal network by the host plant and contribution to the below ground organic carbon pool.

An increase in the carbon supplied by the plant to the AM fungi increases the uptake of phosphorus and the transfer of phosphorus from fungi to plant. Phosphorus uptake and transfer is also lowered when the photosynthate supplied to the fungi is decreased. Species of AMF differ in their abilities to supply the plant with phosphorus. In some cases arbuscular mycorrhizae are poor symbionts providing little phosphorus while taking relatively high amounts of carbon.

The benefit of mycorrhizae to plants is mainly attributed to increased uptake of nutrients, especially phosphorus. This increase in uptake may be due to increase surface area of soil contact, increased movement of nutrients into mycorrhizae, a modification of the root environment and increased storage. Mycorrhizal can be much more efficient than plant roots at taking up phosphorus. Phosphorus travels to the root or via diffusion and hyphae reduce the distance required for diffusion thus increasing uptake. The rate of inflow of phosphorus into mycorrhizae can be up to six times that of the root hairs. In some cases the role of phosphorus uptake can be completely taken over by the mycorrhizal network and all of the plant's phosphorus may be of hyphal origin.

The available phosphorus concentration in the root zone can be increased by mycorrhizal activity. Mycorrhizae lower the rhizosphere pH due to selective uptake of NH_4 (ammonium-ions) and release of H ions. Decreased soil pH increases the solubility of phosphorus precipitates. The hyphal uptake of NH_4 also increases the flow of nitrogen to the plant as NH_4 is adsorbed to the soil's inner surfaces and must be taken up by diffusion.

ECOLOGY

Habitat

Arbuscular mycorrhizal fungi are most frequent in plants growing on mineral soils. The populations of AM fungi is greatest in plant communities with high diversity such as tropical rainforests and temperate grasslands where they have many potential host plants and can take advantage of their ability to colonise a broad host range. There is a lower incidence of mycorrhizal colonisation in very arid or nutrient rich soils. Mycorrhizas have been observed in aquatic habitats; however, waterlogged soils have been show to decrease colonisation in some species.

Host Range and Specificity

The specificity, host range and degree of colonisation of mycorrhizal fungi is difficult to analyse in the field due to the complexity of interactions between the fungi within a root and within the system. There is no clear evidence that arbuscular mycorrhizal fungi exhibit specificity for colonisation of potential AM host plant species as do fungal pathogens for their host plants. This may be due to the opposite selective pressure involved. In parasitic relations, the host plant benefits from mutations which prevent colonisation whereas in a symbiotic relationship, the plant benefits from mutation that allow for colonisation by AMF. However, plant species differ in the extent and dependence on colonization by certain AM fungi and some plants may be facultative mycotrophs while others may be obligate mycotrophs.

The ability of the same AM fungi to colonise many species of plants has ecological implications. Plants of different species can be linked underground to a common mycelial network. One plant may provide the photosynthate carbon for the establishment of the mycelial network which another plant of a different species can utilise for mineral uptake. This implies that arbuscular mycorrhizae are able to balance below ground intra- and inter-specific plant interactions.

Rhizosphere Ecology

The rhizosphere is the soil zone in the immediate vicinity of a root system. Arbuscular mycorrhizal symbiosis affects the community and diversity of other organisms in soil. This can be directly seen by the release of exudates, or indirectly by a change in the plant species and plant exudates type and amount. Mycorrhizae diversity has been shown to increase plant species diversity as the potential number of associations increases. Dominant arbuscular mycorrhizal fungi can prevent the invasion of non-mycorrhizal plants on land where they have established symbiosis and promote their mycorrhizal host. AM fungi release a diffusional factor, known as the myc factor, which activates the nodulation factor's inducible gene mtENOD11. This is the same gene involved in establishing symbiosis with the nitrogen fixing bacteria, rhizobium. When rhizobium bacteria are present in soil, mycorrhizal colonization is increased due to an increase in the concentration of chemical signals involved in the establishment of symbiosis. Effective mycorrhizal colonisation can also increase the nodulations and symbiotic nitrogen fixation in mycorrhizal legumes.

The extent of arbuscular mycorrhizal colonisation and species affects the bacterial population in the rhizosphere. Bacterial species differ in their abilities to compete for carbon compound root exudates. A change in the amount or composition of root exudates and fungal exudates due to the existing AM mycorrhizal colonisation determines the diversity and abundance of the bacterial community in the rhizosphere.

The influence of AM fungi on plant root and shoot growth may also have indirect effect on the rhizosphere bacteria. AMF contributes a substantial amount of carbon to the rhizosphere through the growth and degeneration of the hyphal network. There is also evidence that AM fungi may play an important role on mediating the plant species' specific effect on the bacterial composition of the rhizosphere.

PHYTOREMEDIATION

The use of arbuscular mycorrhizal fungi in ecological restoration projects has been shown to enable host plant establishment on degraded soil and improve soil quality and health (phytoremediation).Disturbance of native plant communities in desertification-threatened areas is often followed by degradation of physical and biological soil properties, soil structure, nutrient availability and organic matter.When restoring disturbed land it is essential to not only replace the above ground vegetation but also the biological and physical soil properties.

A relatively new approach to restore land and protect against desertification is to inoculate the soil with arbuscular mycorrhizal fungi with the reintroduction of vegetation. A long term study demonstrated that a significantly greater long term improvement in soil's quality parameters was attained when soil was inoculated with a mixture of indigenous arbuscular mycorrhizal fungi species compared to the non inoculated soil and soil inoculated with a single exotic species of AM fungi. The benefits observed were an increased plant growth and soil nitrogen content, higher soil organic matter content and soil aggregation. The improvements were attributed to the higher legume nodulation in the presence of AMF, better water infiltration and soil aeration due to soil aggregation.Inoculation with native AM fungi increased plant uptake of phosphorus, improving plant growth and health. The results support the use of AM fungi as a biological tool in the restoration of biotopes to self-sustaining ecosystems.

AGRICULTURE

Many modern agronomic practices are disruptive to mycorrhizal symbiosis. There is great potential for low input agriculture to manage the system in a way that promotes mycorrhizal symbiosis. Conventional agriculture practices, such as tillage, heavy fertilizers and fungicides, poor crop rotations and selection for plants which survive these conditions, hinder the ability of plants to form symbiosis with arbuscular mycorrhizal fungi.

Most agricultural crops can perform better and are more productive when well colonised by AM fungi. AM symbiosis increases the phosphorus and micronutrient uptake and growth of their plant host. Management of AM fungi is especially important for organic and low input agriculture systems where soil phosphorus is generally low, although all agroecosystems can benefit by promoting arbuscular mycorrhizae establishment.

Some crops that are poor at seeking out nutrients in soil are very dependent on AM fungi for phosphorus uptake. For example flax, which has poor chemotaxic ability, is highly dependent on AM mediated phosphorus uptake at low and intermediate soil phosphorus concentrations. Proper management of AMF in the agroecosystems can improve the quality of soil and the productivity of the land. Agricultural practices such as reduced tillage, low phosphorus fertilizer usage and perennialised cropping systems promote functional mycorrhizal symbiosis.

Tillage

Tillage reduces the inoculation potential of the soil and the efficacy of mycorrhizaes by disrupting the extraradical hyphal network. By breaking apart the soil macro structure the hyphal network is rendered non-infective. The disruption of the hyphal network decreases the absorptive abilities of the mycorrhizae because the surface area spanned by the hyphae is greatly reduced. This in turn lowers the phosphorus input to the plants which are connected to the hyphal network.

In reduced tillage system heavy phosphorus fertilizer input may not be required as compared to heavy tillage systems. This is due to the increase in mycorrhizal network which allows mycorrhizae to provide the plant with sufficient phosphorus.

Phosphorus Fertilizer

The benefits of AMF are greatest in systems where inputs are low. Heavy usage of phosphorus fertilizer can inhibit mycorrhizal colonisation and growth. As soil's phosphorus levels available to the plants increases, the amount of phosphorus also increases in the plant's tissues, and carbon drain on the plant by the AM fungi symbiosis become non-beneficial to the plant. A decrease in mycorrhizal colonisation due to high soil phosphorus levels can lead to plant deficiencies in other micronutrients that have mycorrhizal mediated uptake such as copper.

Perennialised Cropping Systems

Cover crops are grown in the fall, winter and spring, covering the soil during periods when it would commonly be left without a cover of growing plants. Mycorrhizal cover crops can be used to improve the mycorrhizal inoculum potential and hyphal network. Since AM fungi are biotrophic, they are dependent on plants for the growth of their hyphal networks. Growing a cover crop extends the time for AM growth into the autumn, winter and spring. Promotion of hyphal growth creates a more extensive hyphal network.

The mycorrhizal colonisation increase found in cover crops systems may be largely attributed to an increase in the extraradical hyphal network which can colonise the roots of the new crop. The extraradical mycelia are able to survive the winter providing rapid spring colonisation and early season symbiosis. This early symbiosis allows plants to tap into the well established hyphal network and be supplied with adequate phosphorus nutrition during early growth which greatly improves the crop yield.

Soil Quality

Restoration of native AM fungi increases the success of ecological restoration project and the rapidity of soil recovery. There is evidence that this enhancement of soil aggregate stability is due to the production of a soil protein known as glomalin.

Glomalin related soil proteins (GRSP) have been identified using a monoclonal antibody raised against crushed AMF spores. It is defined by its extraction conditions and reaction with the antibody Mab32B11. There is other circumstantial evidence that glomalin is of AM fungal origin. When AM fungi are eliminated from soil through incubation of soil without host plants the concentration of GRSP declines. A similar decline in GRSP have also been observed in incubated soils from forested, afforested and agricultural land and grasslands treated with fungicide. Glomalin is hypothesised to improve soil aggregate water stability and decrease soil erosion. A strong correlation has been found between GRSP and soil aggregate water stability in a wide variety of soils where organic material is the main binding agent, although the mechanism is not known. The protein glomalin has not yet been isolated and described, and the link between glomalin, GRSP and arbuscular mycorrhizal fungi is not yet clear.

BENEFITS OF MYCORRHIZAL BIOFERTILIZER

Mycorrhiza plays a very important role on enhancing the plant growth and yield due to an increase supply of phosphorus to the host plant. Mycorrhizal plants can absorb and accumulate several times more phosphate from soil or solution than non–mycorrhizal plants. Plants inoculated with endomycorrhizahave been shown to be more resistant to some root diseases. Mycorrhiza increase root surface area for water and nutrients uptake. The use of mycorrhizal biofertilizer helps to improve higher branching of plant roots, and the mycorrhizal hyphae grow from the root to soil enabling the plant roots to contact with wider area of soil surface, hence, increasing the absorbing area for water and nutrients absorption of the plant root system. Therefore, plants with mycorrhizal association will have higher efficiency for nutrients absorption, such as nitrogen, phosphorus, potassium, calcium, magnesium, zinc, and copper; and also increase plant resistance to drought. Benefits of mycorrhizal biofertiliser can be seemed as follows:

- Allow plants to take up nutrients in unavailable forms or nutrients that are fixed to soil. Some plant nutrients, especially phosphorus, are elements that dissolve were in water in neutral soil. In the extreme acidic or basic soil, phosphorus is usually bound to iron, aluminum, calcium, or magnesium, leading to water insolubility, which is not useful for plants. Mycorrhiza plays an important role in phosphorus absorption for plant via cell wall of mycorrhiza to the cell wall of plant root. In addition, mycorrhiza help to absorb other organic substances that are not fully soluble for plants to use, and also help to absorb and dissolve other nutrients for plants by storage in the root it is associated with.

- Enhance plant growth, improve crop yield, and increase income for the farmers. Arising from improved water and essential nutrients absorption for plant growth by mycorrhiza, it leads to improvement in plant photosynthesis, nutrients translocation, and plant metabolism processes. Therefore, the plant has better growth and yield, reduce the use of chemical fertilizer, sometimes up to half of the suggested amount, which in turn increases income for the farmers. As in the trial involving mycorrhizal biofertilizer on asparagus it was observed that, when the farmers used suggested amount of chemical fertilizer together with mycorrhizal biofertilizer, it was found that the crop yield improved by more than 50%, and the farmers' income increased 61% higher than when chemical fertilizer alone was used.It can be used together with other agricultural chemicals. Mycorrhiza are endurable to several chemical substances; for example; pesticide such as endrin, chlordane, methyl parathion, methomyl carbofuran; herbicide such as glyphosate, fuazifopbutyl; chemical agents for plant disease elimination such as captan, benomyl, maneb triforine, mancozed and zineb.

ISOLATION OF ARBUSCULAR MYCORRHIZAL FUNGI

Taxonomy of AM Fungi

AM fungi show the peculiar characteristics in morphology and physiology. Spores of AM fungi are generally formed in soil and their sizes (50-500 ìm in diameter) are much larger than those of other fungi. There is no septum in their hyphae. No sexual growth-phase has been observed. Spores germinate when they are under favourable conditions, extend their hyphae and colonised plant roots. The fungi penetrate the hyphae into cortex layer of roots and form the hyphal organs, "vesicles" and "arbuscules" which are characteristics to AM fungi. AM fungi belonging to Gigasporaceae are known not to form vesicles. Colonisation on plant roots is essential for proliferation of AM fungi. AM fungi are thus recognised as obligate symbiotic fungi. The interaction between AM fungi and plants is generally mutualism based upon nutrient exchange.

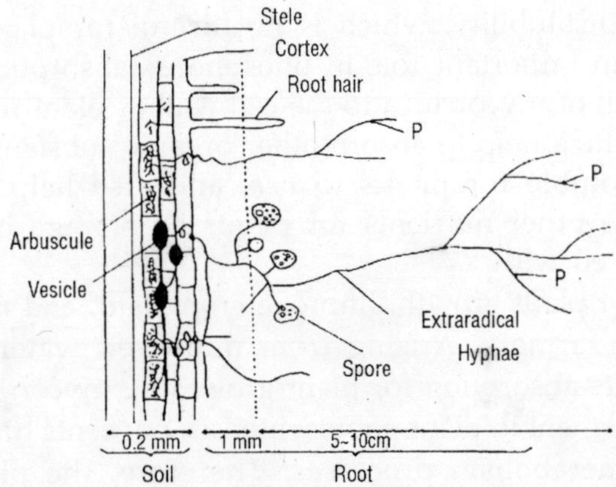

Fig:1. Schematic Picture of Arbusucular Mycorrhizal Fungi Colonising Roots and Their Hyphal Extension into Soil

Because of morphological characteristics such as no hyphal septum, AM fungi had long been recognised as a member of Zygomycota. Recent molecular phylogenetic studies showed that Zygomycota is poly-phyletic and that AM fungi should be separated from other Zygomycota.

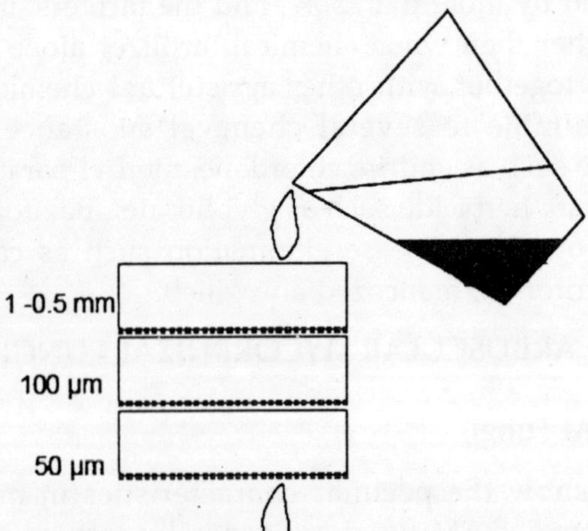

Fig:2. Taxonomy of Arbuscukar Mycorrhizal Fungi

A new Phylum Glomeromycota has been proposed for AM fungi. This classification is mainly based upon the sequence data of rRNA gene. However, some new genera have been raised with relatively small numbers of isolates, so further study may revise the present classification system. Although there is a recent trend that the sequence data of AM fungi is over-emphasised for the identification, the conventional morphological observation is still important and should not be neglected for identification.

Observation of Arbuscular Mycorrhizal Fungi in Roots

Arbuscular mycorrhizal fungal structure in roots is usually not observed without appropriate staining. Freshly collected root samples should be washed gently and be free from soil particles. Ultrasonic treatment is effective to disperse soil particles closely adhered to roots. Roots are treated with 10 per cent KOH solution for 30 min to 1-2 hours in a hot bath, depending on thickness of root structure.

Treated roots are washed with water and treated with 2% HCl solution. Acidified root samples are stained with 0.05% trypan blue (or acid fuchsin) in lactic acid for 10-15 min. in a hot bath or for a few hours without heating.

The roots are destained with lactic acid or lacto-glycerol and are now ready for microscopic observation. The stained roots may be observed first under a dissecting microscope with transmitted illumination and then observed under a compound microscope. Fungal structures are stained and can be easily recognised.

Isolation of Spores from Soils and Their Observation for Identification

Spores of AM fungi in soil can be collected by the wet sieving method. The gravity of spores is a little lighter than that of soil particles. Successive decantation of soil suspension followed by sieving with fine mesh can concentrate the spores from soil. Since the spores are globular or sub–globular in 50–500 ìm in diameter, they, in sievings can be recognised under a dissecting microscope. Equipments:

- *Sieve:* Sieves with various mesh size. At least the following mesh size are required; 1 mm, 100 ìm and 50 ìm. Other sizes such as 500 ìm and 250 ìm are preferable. Stainless steel sieves are commercially available. However, it is possible to make a plastic sieve with PVP tubes and nylon mesh by yourself.
- *Fine glass pipettes:* Tip of disposable glass Pasteur pipette (1 ml) is softened with flame of gas burner and is sharpened. Various sizes of tips fitting to sizes of spores can be prepared.
- *Dissecting microscope:* Stereoscopic zoom microscope with bifurcated illuminator of fibre arm is advisable. Transmitted illumination system is also needed.
- *Compound microscope:* Biological compound microscope is needed. Nomarsky's DIC illumination system is advisable.

Ten to 50 g of freshly collected soil sample is put into 1 to 2 litres of plastic beakers. Usually rhizosphere soils are rich in AM fungal spores. Beaker size can be changed depending on the soil sample size. Soil is suspended with about 500 ml to 1 litre of tap water. Soil macro-aggregates should be crushed with hand. After 10-30 seconds of settling down of soil particles, the upper layer of soil suspension is poured into the sieving. The procedure should be repeated until the upper layer of soil suspension is transparent.

The sievings on the fine mesh is collected into a small beaker and dispersed with ultra sonication. Weak sonication (i.e., 30W 30 sec) is enough, and strong sonication may destroy fungal spores. Then the dispersed sample is again passed through the sieve. Depending on toughness of soil aggregate, the sonication process can be repeated. Usually AM fungal spores are collected on 100 ìm. Some small spores are on 50 ìm. To collect large spores such as *Gigaspora margarita*, 250 ìm sieve is efficient.

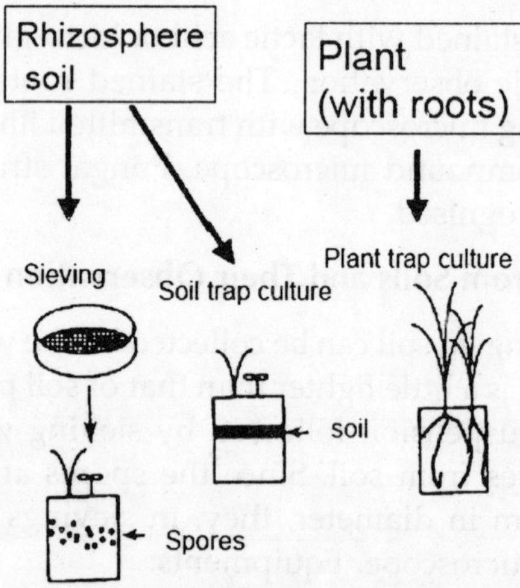

Fig:3. Sieving of Soil Suspension

Spores of AM fungi have characteristic shapes and colours, and so it is not difficult to discriminate the spores in organic debris collected on the sieves. However, it is recommended that those who have not yet observed the spores before should learn from the experts how the spores look like. In the soil such as grassland soil rich in organic debris, it may be hard to find the spores hidden by the debris. In such a case, sucrose density centrifugation technique is often used to separate spores from the organic debris.

Morphological Observation of Spores for Identification

Morphology of spores is a basis for identification of AM fungi, because the hyphae and the organs such as arbuscules and vesicles are not specific to species. Spores collected from soil often deteriorate so that they may be used only for tentative identification at genus level. For detailed observation, culturing the target AM fungus is required, and the spores from pot culture should be used. At least 30-50 spores from the same morphological spore type should be observed, and more observation is recommended.

Observation of Intact Spores under Dissecting Microscope

Spores collected from soil or culturing medium are put in a watchglass or a small Petri dish, and their shape, colour and the attachment to spores are observed. Spores should be classified into each spore type based upon morphology. For each spore type, detailed observation is conducted. For colour description, standard colour chart such as soil colour chart or "colour chart of glomalean fungi" should be used. The colour chart should be under the same illumination as used for spore observation, because the colour itself is greatly affected by the characteristics of illumination. It should be reminded that hyphal attachments such as sporiferous saccule for Acaulosporaceae and subtenting hyphae for Gigasporaceae are often lost during collection of spores from soil.

Observation of Spores Mounted on Slide Glass under a Compound Microscope

Spores are mounted with polyvinyl lactoglycerol (PVLG) on a slide glass. Several slides should be made. These are for intact spores mounted with PVLG, for crushed spores mounted with PVLG, for spores mounted with PVLG containing Meltzer's reagent.

These characteristics may be helpful to identify genus of the target fungus. However, some genera such as *Archaeospora* needs not only these morphological characteristics but also sequence data. For species identification, the characteristics should be compared with those in the species description in the original reference. Many species description and pictures are also available in INVAM website. Detailed information for morphological observation is available in review by Morton and his website.

Species identification without enough expertise may cause trouble in scientific society, so those who wish to identify species of AM fungi should consult the experts in AM fungal taxonomy. For spore size, at least 40-50 spores should be examined, and more examination is preferable. PVLG (poly-vinyl lacto-glycerol): Polyvinyl alcohol (polymerization 1000-1500), 1.66 g, is dissolved in 10 ml of deionised water. Complete dissolution may need 6 hours at 80°C. The dissolved polyvinyl alcohol is mixed with 10 ml of lactic acid and 1 ml of glycerol. It can be used more than a day after preparation.

Culturing AM Fungi

AM fungi need the symbiotic association with plants for proliferation. Therefore, culturing AM fungi is to inoculate AM fungi to host plant and to grow the inoculated plant. For the AM fungal inoculum, spores collected form soil can be used. However, spores in soil are not always active in colonising plants. Therefore, trapping culture is often employed. Soil or sieving of soil is used as inoculum (Soil Trap Culture).

To isolate AM fungi colonising roots, mycorrhizal plants collected from field can be transplanted to potting medium as Plant Trap Culture. Potting medium: Sterile soil or soil-sand mixture is usually used. We prefer to use commercially available "Akadama-tsuchi" which is collected from subsoil of volcanic ash soil and is prepared for horticulture use. Various potting materials for horticulture can be also used.

However, the materials for potting medium should be low in available phosphate and preferably not rich in organic matter. In some cases the fungi isolated from some specific soils may need the specific soil properties for their growth. Host plant: Various mycotrophic plants can be used: leguminous species and grass species (i.e., *Lolium* spp., *Paspalum notatum*), and other herbaceous species (i.e., *Plantago* spp.).

Onion and leek (*Allium* spp.) are also good hosts. AM fungi generally do not show host specificity but some species show host preference. Therefore, the plant species from which the target AM fungus is isolated can be used as a host plant. Growth conditions: Any conditions, which support good growth of host plants, are acceptable. To avoid contamination, a growth chamber is preferable. If greenhouse is used, it should be kept clean. It should be reminded that cross-contamination or contamination from dust is inevitable under open-air conditions, even in growth chamber.

To prevent cross-contamination from other pot culture in the same chamber, use of plastic bag is advisable. Single spore isolation: To purify an isolated fungus, single spore isolation is needed. Even if the spores are morphologically identical, it often contains contaminants whose morphology is very similar. Successive pot culture of such multispore isolates would cause unexpected outbreak of the contaminant. Furthermore, even if the culture contains only one species, it may be composed of genetically diverse populations. For such a genetic studies or population genetics, the purification through single spore isolation is essential.

For single spore isolation, no specific equipment is needed. For efficient handing, two sets of dissecting microscopes are place side by side. One microscope is for picking up single spore from spores in a dish. Another is for inoculation of a spore on roots. Seedling placed in the pot is placed under another microscope.

Under the first microscope, single spore is picked up and transferred under the second microscope. Under the second microscope, the spore is placed on fine roots or root tip of the seedling. If culture is successful, the detailed morphological observation is required. Potting medium can be dried by stopping watering to the pot. After the host plant wilt, the dried soil containing spores can be stored for a year at 4-5°C. It is advisable that the isolated fungi are re-cultured every year.

DNA EXTRACTION FROM SPORES

Molecular phylogenetic information is essential for taxonomy of AM fungi. Sequence data for conserved genes such as rRNA is obtained by PCR amplification followed by sequencing with DNA extracted from spores. Many protocols for DNA extraction are reported. One of the protocols which we have used is described. Cleaning of spores: Clean spores are collected with tweezers or fine Pasture pipette. Spores should wash several times with sterile water with sonication for 10-20 seconds. If spores show water repellency, surfactant such as Tween 80 can be used. Microplate with 6 or 12 wells is convenient for successive washing. Crushing spores: We use a cap of Eppendorf tube. The cap is cut from the tube and placed upside down on the stage of a dissecting microscope. A spore or spores are put into the cap and crushed thoroughly with a micro-pestle or any fine rod. The tube itself is put on the cap with crushed spores.

Then the tube is centrifuged for while for spinning down the reagent with crushed spores from the cap into the tube. Extraction of DNA: Follow the instruction provided by manufacture. Further purification of DNA with ethanol/chloroform precipitation is sometimes needed to remove inhibitor for PCR amplification. PCR amplification and DNA sequencing: Conditions of PCR amplification depend on the primers you will use. AM fungi are multi-nuclear organisms and often show polymorphisms in their sequence. It is advisable to sequence several clones from the target fungus and to check the phylogenetic position of the sequence by constructing phylogentic tree if the target genes are located within a reasonable clade of the tree.

Chapter 4

Soil Chemistry

SOIL

Gardeners know that you cannot grow vegetables just anywhere. The quality of vegetables grown depends on the condition of soil. Soil is the top layer of the Earth's surface that is suitable for plant growth. Soil quality is a major concern for Prince Edward Island farmers because of its importance to agriculture. Although it is often referred to as dirt or ground, there is a lot more to soil than meets the eye.

PH

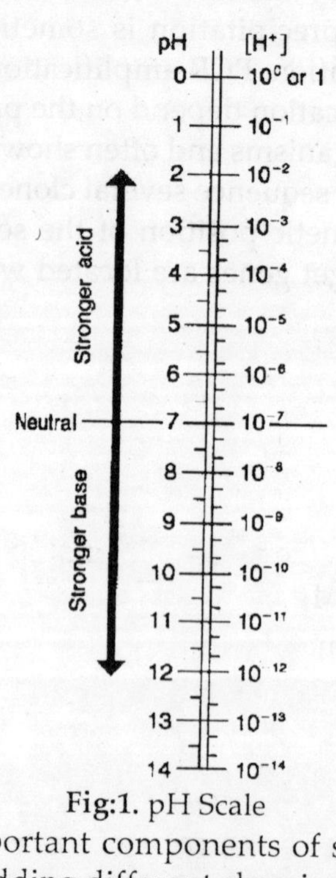

Fig:1. pH Scale

One of the most important components of soil is the pH. The pH of soil can be modified by adding different chemicals. Soil pH indicates how acid or alkaline the soil is. The pH scale ranges from 0 to 14. Any substance with a pH near the lower end of the scale is very acidic. Substances in the upper range of the scale have a high alkalinity or are very basic. The pH of a soil is crucial because crops grow best in a narrow pH range which can vary among crops. For example, blueberries and a few types of flowers grow best when the pH is 5.5 or less. Potatoes, a more familiar crop, grow best with a soil pH range of 5.5 to 6.0. Most garden vegetables, shrubs, trees and lawns grow best when the soil pH is over 6.0 or 6.5. The range between 5.5 and 7.5 is favorable for two reasons It allows sufficient microorganisms to break down organic matter. It is also the best range for nutrient availability. Areas that were formerly covered by trees, such as Prince Edward Island, develop acidic soils. This helps explain how Prince Edward Island farmers can grow the best potatoes in the world.

LIMING

When farmers originally cleared the lands on the Island, the soil quality was adequate for growing potatoes. However, potatoes were not the only crop at that time. Therefore, farmers needed a way to increase the pH of the soil to make it suitable for other crops. The pH of soil can be increased by liming. This is why people sometimes spread white powder on their lawns or gardens. This white powder is lime. Calcitic limestone (CaCO) provides a good 3 source of Calcium (Ca) and helps neutralise soil acidity. Dolomitic limestone functions similarly but also adds magnesium (Mg). The best limestone will have the greatest calcium and magnesium content and will be ground into very tiny particles. The smaller particles allow the limestone to correct soil acidity more rapidly. The chemistry to liming is quite simple. Hydrogen ions (H^+) are attracted to soil and organic material which have a negative charge. When lime is applied, these hydrogen ions are exchanged for calcium or magnesium (Ca^{2+} or Mg^{2+}) ions which have a greater positive charge. This helps to neutralise the acidity of soil. The free hydrogen ions are taken out of solution. This also helps to increase the pH. This reaction demonstrates the process of liming:

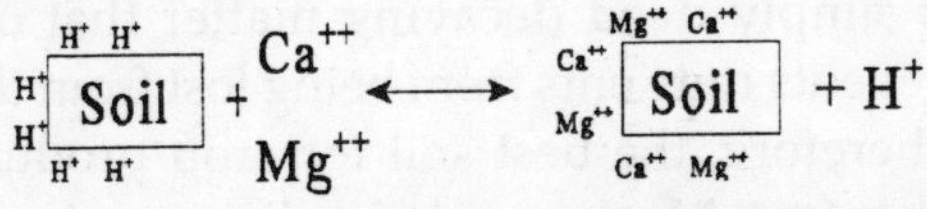

In some cases, the soil may have very high pH and need to be made more acidic. This can be done by using sulfur, aluminum sulfate, or ammonium sulfate.

NUTRIENTS

Each year the soil undergoes a series of cycles in which materials are added and then taken away. Organic matter and nutrients, in various forms, are constantly being added to soil. Nutrients are the minerals required by plants to survive. It is very important that plants receive all the required nutrients. There are a total of sixteen elements required for plant growth. Each is required in different amounts. The most important nutrients are called macronutrients. Nutrients which are essential, but only needed in small quantities are called micronutrients. Carbon, Hydrogen, and Oxygen are the big three macronutrients.

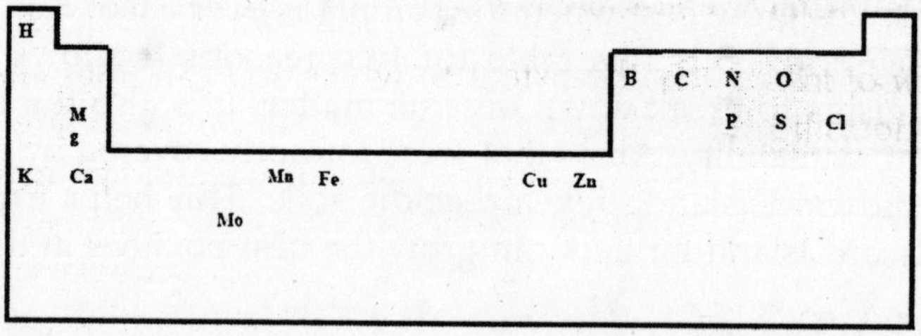

Fig:2. The Sixteen Essential Nutrients Indicating Their Position on the Periodic table. The Macronutrients are Highlighted

These are obtained in almost unlimited amounts from the atmosphere and from the water around the plant. The other macronutrients are nitrogen (N), phosphorus (P), potassium (K), calcium (Ca), magnesium (Mg), and sulfur (S). These nutrients should be available for plant uptake from soil. Micronutrients are obtained from soil. They are boron (B), copper (Cu), chlorine (Cl), iron (Fe), manganese (Mn), molybdenum (Mo), and zinc (Zn). They are all essential because the absence of any one of these will cause the plant to grow poorly or develop disease.

ORGANIC MATTER

Many pleasant drives in the country have been affected when the passengers of a car are hit with an unpleasant, but familiar odour. Many people complain about the smell without questioning why it exists. There is actually a very good reason for this practice. Although they are often given more unpleasant names, these animal wastes are known as organic fertilizers. These fertilizers have a very high content of organic matter. Organic matter is simply dead decaying matter that originated from a living source. It prevents nutrients from being lost from the soil by binding these nutrients. Therefore, the best soil for crop production will have a very high organic content. Most organic fertilizer originates from livestock such as cows, pigs, and poultry. Compost is also an organic fertilizer. Compost can be made up of grass clippings, table scraps, ashes, seaweed, and many other types of food products.

Organic fertilizers contain high levels of Nitrogen and moderate levels of Phosphorus and Potassium. The nutrient content of organic fertilizer can vary according to the animal that produced it (agriscience text). The process of spreading organic fertilizers gives farmers the opportunity to rid themselves of accumulated livestock waste. It also provides farmers with a free source of fertilizer which is sometimes sufficient to meet the needs of the desired crop. Organic fertilizers are also less harmful to the environment. This may be one of the first recycling practices that ever developed.

CATION-EXCHANGE CAPACITY

The CEC measures the extent to which soil can hold and exchange plant nutrient cations. The ability of soil to hold positively charged nutrients from being leached and lost from soil is important to maintaining soil fertility. Clay and organic matter have a negative charge. They allow the soil to hold these nutrient cations due to the attraction of charges. Soils with high clay or organic matter content will have a higher CEC. Sandy soils tend to have a lower CEC.

BUFFERING CAPACITY

This is the ability to withstand rapid pH fluctuation. The greater the buffering capacity, the greater the quantity of acid or base which must be used to alter the pH. Soil types having low buffering capacities include sandy soils with little clay or organic matter. Soils with a higher buffering capacity would have large quantities of mineral clay and organic matter. Therefore, a thick rich soil with a high buffering capacity would require more lime in order to raise the pH.

FERTILIZER

Fertilizers can replenish the mineral nutrients depleted from soil by natural means or during crop removal. Most fertilizers contain nitrogen (N), phosphorus (P), and potassium (K). The proportions of these elements are called the fertilizer grade. The fertilizer grade on a bag gives the percentages of each mineral by weight. For example, a bag of fertilizer labeled 10-10-10 contains 10 per cent N, 10 per cent P, and 10 per cent K. In the common 80 pound bag of this grade, you would have 8 pounds of each nutrient. There are a variety of different grades of fertilizer. The cost varies according to the contents. The most appropriate grade of fertilizer depends on the desired crop, and the condition of the soil where the crop will be planted.

SOIL FORMATION AND CLASSIFICATION

Soil Forming Factors

Parent Material

Few soils weather directly from the underlying rocks. These

"residual" soils have the same general chemistry as the original rocks. More commonly, soils form in materials that have moved in from elsewhere.

Materials may have moved many miles or only a few feet. Windblown "loess" is common in the Midwest. It buries "glacial till" in many areas. Glacial till is material ground up and moved by a glacier. The material in which soils form is called "parent material". In the lower part of the soils, these materials may be relatively unchanged from when they were deposited by moving water, ice, or wind.

Sediments along rivers have different textures, depending on whether the stream moves quickly or slowly. Fast-moving water leaves gravel, rocks, and sand. Slow-moving water and lakes leave fine textured material (clay and silt) when sediments in the water settle out.

Climate

Soils vary, depending on the climate. Temperature and moisture amounts cause different patterns of weathering and leaching. Wind redistributes sand and other particles especially in arid regions. The amount, intensity, timing, and kind of precipitation influence soil formation. Seasonal and daily changes in temperature affect moisture effectiveness, biological activity, rates of chemical reactions, and kinds of vegetation.

Topography

Slope and aspect affect the moisture and temperature of soil. Steep slopes facing the sun are warmer, just like the south-facing side of a house. Steep soils may be eroded and lose their topsoil as they form. Thus, they may be thinner than the more nearly level soils that receive deposits from areas upslope. Deeper, darker colored soils may be expected on the bottom land.

Biological Factors

Plants, animals, microorganisms, and humans affect soil formation. Animals and microorganisms mix soils and form burrows and pores. Plant roots open channels in the soils. Different types of roots have different effects on soils. Grass roots are "fibrous" near the soil surface and easily decompose, adding organic matter. Taproots open pathways through dense layers. Microorganisms affect chemical exchanges between roots and soil. Humans can mix the soil so extensively that the soil material is again considered parent material.

The native vegetation depends on climate, topography, and biological factors plus many soil factors such as soil density, depth, chemistry, temperature, and moisture. Leaves from plants fall to the surface and decompose on the soil. Organisms decompose these leaves and mix them with the upper part of the soil. Trees and shrubs have large roots that may grow to considerable depths.

Time

Time for all these factors to interact with the soil is also a factor. Over time, soils exhibit features that reflect the other forming factors. Soil formation processes are continuous. Recently deposited material, such as the deposition from a flood, exhibits no features from soil development activities. The previous soil surface and underlying horizons become buried. The time clock resets for these soils. Terraces above the active floodplain, while genetically similar to the floodplain, are older land surfaces and exhibit more development features.

These soil forming factors continue to affect soils even on "stable" landscapes. Materials are deposited on their surface, and materials are blown or washed away from the surface. Additions, removals, and alterations are slow or rapid, depending on climate, landscape position, and biological activity.

When mapping soils, a soil scientist looks for areas with similar soil-forming factors to find similar soils. The colours, texture, structure, and other properties are described. Soils with the same kind of properties are given taxonomic names.

A common soil in the Midwest reflects the temperate, humid climate and native prairie vegetation with a thick, nearly black surface layer. This layer is high in organic matter from decomposing grass. It is called a "mollic epipedon". It is one of several types of surface horizons that we call "epipedons".

Soils in the desert commonly have an "ochric" epipedon that is light coloured and low in organic matter. Subsurface horizons also are used in soil classification. Many forested areas have a subsurface horizon with an accumulation of clay called an "argillic" horizon.

Soil Development

Soil forms from a complex interaction between earth materials, climate, and organisms acting over time. The brightly coloured soils of the humid tropics reflect the intense chemical reactions occurring in warm climates. The fertile prairie soils of the American Midwest evolved from the nutrient-rich organic matter left by decaying grasses. Regardless of soil characteristics, the whole process starts with the breakdown of earth material.

Fig:3. Soil Profile in Central Iowa

Weathering

Weathering refers to processes that physically breakdown and chemically alter earth material. *Physical weathering,* also known as *mechanical weathering,* is the breakdown of large pieces of earth material into smaller ones.

Think of physical weathering as the *disintegration* of rock without changing its chemical composition. There are many ways earth material can be physically weathered. When water freezes in rock crevices it expands creating stress in the crevice. As the stress increases, the crevice widens ultimately breaking the rock. Plant roots wedge rocks apart as they grow into rock crevices too. The shrinking and swelling by alternating heating and cooling weakens mineral bonds causing the rock to disintegrate.

A very important result of physical weathering is its impact on the surface area of weathered material. When a block of earth material is broken into several smaller pieces, the amount of exposed surface increases. Examine the diagram below. A block with a width, depth, and height of 1 cm has a total surface area of 6 square centimetres. If we break the block in half in all directions it yields eight smaller pieces all with width, height, and depth of 0.5 cm.

Breaking the block apart creates additional exposed surfaces such that the total surface area is now 12 square centimetres. Having more total exposed surface provides more area upon which chemical reactions can take place to further weather the material. The shape of the pieces also affects the the amount of exposed surface area. Plate-like pieces have more exposed surface area than do block-like pieces.

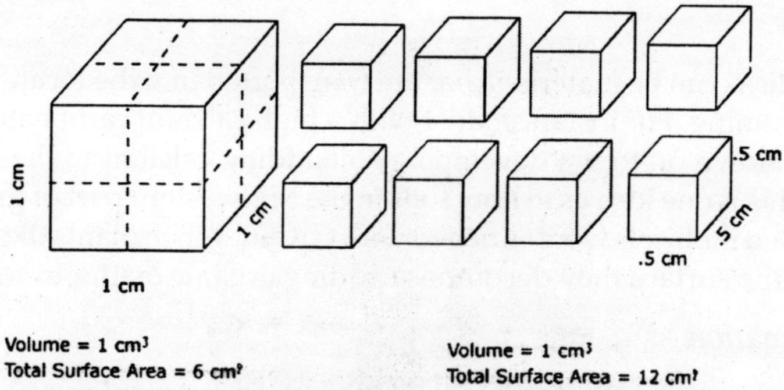

Fig:4. Effect of Physical Weathering on Surface Area

Chemical weathering breaks down earth material by chemical alteration. This usually means adding a substance like water or air to the material. For instance, when oxygen is added to iron bearing minerals, *oxidation* takes places and a loose mantle of iron oxide is created (rust). *Hydrolysis* is an exchange reaction involving minerals and water. Free hydrogen (H^+) and hydroxide $(OH)^-$ ions in water replace mineral ions and drive them into solution. As a result, the mineral's structure is changed into a new form. Hydrolysis is a common process whereby silicate minerals are weathered into a clay mineral. Think of chemical weathering as the decomposition of earth material.

The spatial variation of climate and organisms play a significant role in the weathering of earth materials. Dry locations tend to be dominated by physical weathering and moist places by chemical weathering. The type of earth material available also determines the amount of weathering that might take place. Limestone is easily broken down where abundant rainfall and high temperatures prevail. However, limestone will remain intact in dry locations. The end result of the weathering process is the creation of a *weathered mantle*. The weathered mantle is not yet a soil until it undergoes further change. This involves the addition, transformation, translocation and removal of materials from the weathered mantle to form distinctive soil layers.

Horizon Development Processes

Ever dug into in the soil and noticed how it seems to change colour the deeper you go? Some soils are dark brown near the surface and get lighter in colour as you go deeper. Others display a sandy, light-gray layer near the surface with a reddish layer beneath. These different coloured layers are known as *horizons*. All the horizons taken together comprise the *soil profile*. Soil horizons form as a result of the four horizon development processes, additions, transformation, translocation, and removal.

Additions

Additions can be materials that are transported into the location where a soil is forming. For instance, dust with a high calcium carbonate content could be blown on to the developing soil adding calcium to the evolving profile. This is one idea as to how soils in the Midwestern part of the United States called mollisols were enriched with $CaCO_3$. When plants die or leaves fall on to the surface they decompose adding organic matter to soil.

Transformation

Transformation of the materials added to the developing soil occur by chemical and biological processes acting on them. For instance, leaves falling on surface and plant roots dying beneath may decompose into a dark brown, nutrient-rich material called *humus*. Humus is responsible for the dark brown to black colour of many soils, especially near the top of the soil profile. Iron and aluminum can be oxidized under warm, moist climates. Soil material is constantly being transformed in one way or another.

Translocation

Translocation involves the movement of soil-forming materials through the developing soil profile. Translocation occurs by water running through soil transferring materials from upper to lower portions of the profile. Burrowing animals like earth worms, ants, etc., move soil materials within the profile. Burrowing animals create passage ways through which air and water can travel promoting soil development.

Removal

Removal of soil forming materials means that they are completely removed from the soil profile. Easily dissolved elements like calcium carbonate can be removed from the soil profile under rainy climates.Soil horizons develop in response to the relative importance of each of the above processes. All soils are impacted by the horizon development processes to one degree or another. These processes determine the characteristics or properties of soil in each of the horizons.

SOIL PROPERTIES

Soils can be enormously complex systems of organic and inorganic components. Here, we'll concentrate on a few of the most significant properties, texture, structure, colour, and chemistry.

Soil Texture

Soil texture refers to the relative proportion of sand, silt and clay size particles in a sample of soil. Clay size particles are the smallest being less than 0.002 mm in size. Silt is a medium size particle falling between 0.002 and 0.05 mm in size. The largest particle is sand with diameters between 0.05 for fine sand to 2.0 mm for very coarse sand. Soils that are dominated by clay are called fine textured soils while those dominated by larger particles are referred to as coarse textured soils. Soil scientists group soil textures into *soil texture classes*. A soil texture triangle is used to classify the texture class.The sides of the soil texture triangle are scaled for the percentages of sand, silt, and clay. Clay percentages on the left side of the triangle are read from left to right across the triangle (dashed lines).

Silt runs from the top to the bottom along the right side and is read from the upper right to lower left (light, dotted lines). The percentage of sand increases from right to left along the base of the triangle. Sand is read from the lower right towards the upper left portion of the triangle (bold, solid lines). The boundaries of the soil texture classes are highlighted in blue. The intersection of the three sizes on the triangle give the texture class. For instance, if you have a soil with 20% clay, 60% silt, and 20% sand it falls in the "silt loam" class.

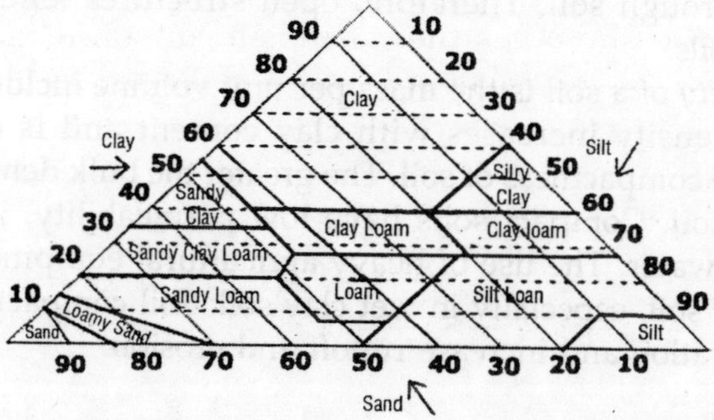

Fig:5. Soil Texture Triangle

Soil texture effects many other properties like structure, chemistry, and most notably, soil porosity, and permeability. *Soil porosity* refers to the amount of pore, or open space between soil particles. Pores are created by the contacts made between irregular shaped soil particles. Fine textured soil has more pore space than coarse textured because you can pack more small particles into a unit volume than larger ones. More particles in a unit volume creates more contacts between the irregular shaped surfaces and hence more pore space. As a result, fine textured clay soils hold more water than coarse textured sandy soils. *Permeability* is the degree of connectivity between soil pores. A highly permeable soil is one in which water runs through it quite readily. Coarse textured soils tend to have large, well-connected pore spaces and hence high permeability.

Soil Structure

Soil structure is the way soil particles aggregate together into what are called peds. Peds come in a variety of shapes depending on the texture, composition, and environment.

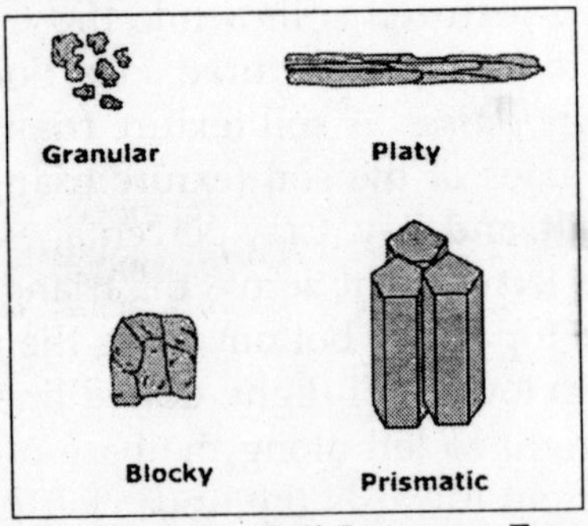

Fig:6. Common Soil Structure Forms

Granular, or crumb structures, look like cookie crumbs. They tend to form an open structure that allows water and air to penetrate the soil. *Platy structure* looks like stacks of dinner plates overlaying one another. Platy structure tends to impede the downward movement of water and plant roots through soil. Therefore, open structures tend to be better agricultural soils.

Bulk Density of a soil is the mass per unit volume including the pore space. Bulk density increases with clay content and is considered a measure of the compactness of soil. The greater the bulk density, the more compact the soil. Compact soils have low permeability, inhibiting the movement of water. The use of heavy agricultural equipment can cause compaction of soil, especially in wet clay soil. Soil compaction results in reduced infiltration and increase runoff and erosion.

Soil Chemistry

As plant material dies and decays it adds organic matter in the form of humus to the soil. Humus improves soil moisture retention while affecting soil chemistry. Cations such as calcium, magnesium, sodium, and potassium are attracted and held to humus. These cations are rather weakly held to the humus and can be replaced by metallic ions like iron and aluminum, releasing them into the soil for plants to use. Soils with the ability to absorb and retain exchangeable cations have a high cation-exchange capacity.

Soils with a high cation-exchange capacity are more fertile than those with a low exchange capacity. Hydrogen ion concentration in soil is measured in terms of the *pH scale*. Soil pH ranges from 3 to 10. Pure water has a pH of 7 which is considered neutral, pH values greater than seven are considered *basic or alkaline*, below seven *acidic*. Most good agricultural soils have a pH between 5 and 7. Though acidic soils pose a problem for agriculture due to their lack of nutrients, alkaline soils can pose a problem as well. Alkaline soils may contain appreciable amounts of sodium that exceed the tolerances of plants, contribute to high bulk density and poor soil structure. Alkaline soils are common in semiarid regions.

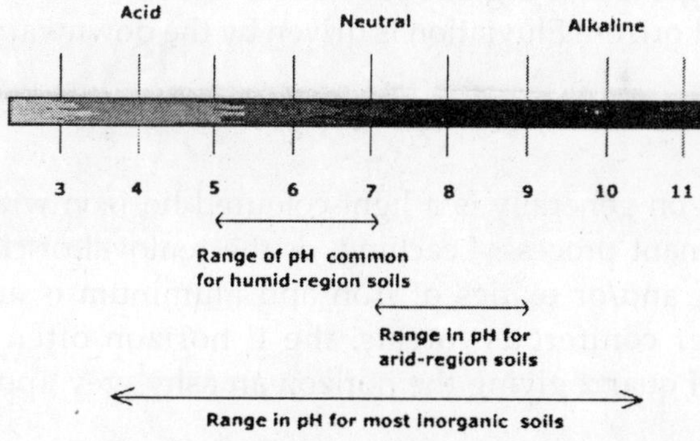

Fig:7. Soil pH

SOIL PROFILES

Soil formation begins first with the break down of rock into regolith. Continued weathering and soil horizon development process leads to the development of a *Soil profile*, the vertical display of soil horizons. Watch the typical progression of a soil profile then read the description below of a generic, fully developed soil.

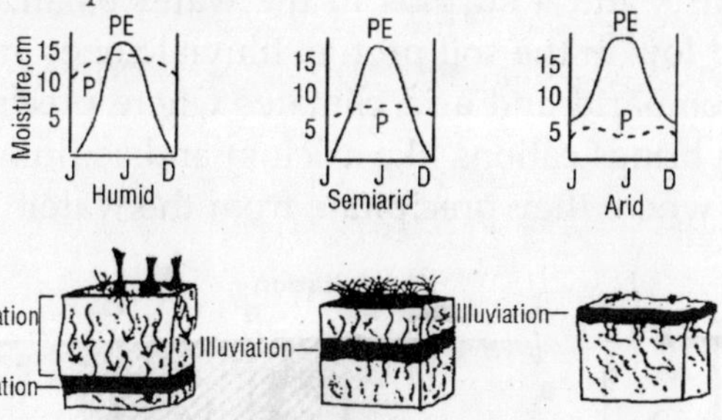

Fig:8. A Typical Soil Profile

O Horizon

At the top of the profile is the O horizon. The O horizon is primarily composed of organic matter. Fresh litter is found at the surface, while at depth all signs of vegetation structure has been destroyed by decomposition. The decomposed organic matter, or humus, enriches the soil with nutrients (nitrogen, potassium, etc.), aids soil structure (acts to bind particles), and enhances soil moisture retention.

A Horizon

Beneath the O horizon is the A horizon. The A horizon marks the beginning of the true mineral soil. In this horizon organic material mixes with inorganic products of weathering. The A horizon typically is dark coloured horizon due to the presence organic matter. *Eluviation*, the removal of inorganic and organic substances from a horizon by leaching occurs in the A horizon. Eluviation is driven by the downward movement of soil water.

E Horizon

The E horizon generally is a light-coloured horizon with eluviation being the dominant process. Leaching, or the removal of clay particles, organic matter, and/or oxides of iron and aluminum is active in this horizon. Under coniferous forests, the E horizon often has a high concentration of quartz giving the horizon an ashy-grey appearance.

B Horizon

Beneath the E horizon lies the B horizon. The B horizon is a zone of *illuviation* where downward moving, especially fine material, is accumulated. The accumulation of fine material leads to the creation of a dense layer in the soil. In some soils the B horizon is enriched with calcium carbonate in the form of nodules or as a layer. This occurs when the carbonate precipitates out of downward moving soil water or from capillary action. Eluviation is significant in humid climates where ample precipitation exists and a surplus in the water balance occurs. Illuvial layers are found low in the soil profile. Illuvial zones are found closer to the surface in semi-arid and arid climates where precipitation is scarce. Capillary action brings cations like calcium and sodium dissolved in soil water upwards where they precipitate from the water.

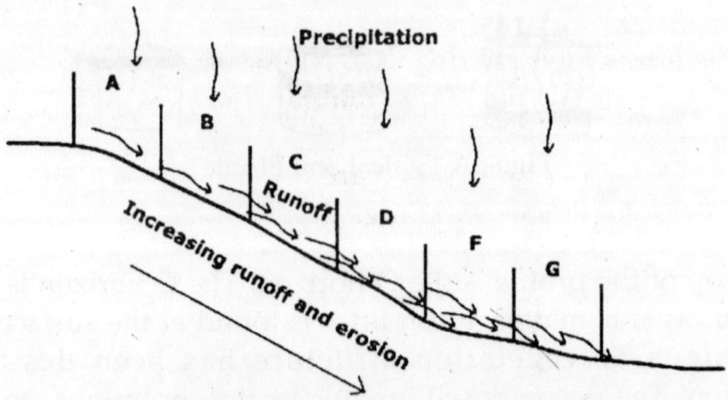

Fig:9. Eluviation And Illuviation Under Humid, Semiarid And Arid Conditions.

C Horizon

The C horizon represents the soil parent material, either created *in situ* or transported into its present location. Beneath the C horizon lies bedrock.The preceding paragraphs describe a generic soil profile, yet not all soils have each one of the horizons, nor are they all the same with respect to thickness composition and structure. Newly formed "immature" soils may only have an O-A-C sequence while older more "mature" soils display the full profile of horizons as described above. The particular compositional, structural and chemical composition of the soil depends on the various factors that influence soil formation.

FACTORS AFFECTING SOIL DEVELOPMENT

Soil are influenced by five separate, yet interacting, factors: parent material, climate, topography, organisms, and time. Soil scientists call these the *factors of soil formation*. These factors give soil profiles their distinctive character.

Parent Material

Soil *parent material* is the material that soil develops from, and may be rock that has decomposed in place, or material that has been deposited by wind, water, or ice. The character and chemical composition of the parent material plays an important role in determining soil properties, especially during the early stages of development. Soils developed on parent material that is coarse grained and composed of minerals resistant to weathering are likely to exhibit coarse grain texture. Fine grain soil develop where the parent material is composed of unstable minerals that readily weather.

Parent material composition has a direct impact on soil chemistry and fertility. Parent materials rich in soluble ions-calcium, magnesium, potassium, and sodium, are easily dissolved in water and made available to plants. Limestone and basaltic lava both have a high content of soluble bases and produce fertile soil in humid climates. If parent materials are low in soluble ions, water moving through the soil removes the bases and substitutes them with hydrogen ions making the soil acidic and unsuitable for agriculture.

Soils developed over sandstone are low in soluble bases and coarse in texture which facilitates leaching. Parent material influence on soil properties tends to decrease with time as it is altered and climate becomes more important.

Climate

Soils tend to show a strong geographical correlation with climate, especially at the global scale. Energy and precipitation strongly influence physical and chemical reactions on parent material. Climate also determines vegetation cover which in turn influences soil development. Precipitation also affects horizon development factors like the translocation of dissolved ions through the soil. As time passes, climate tends to be a prime influence on soil properties while the influence of parent material is less.

Climate, Vegetation and Weathering

Climate affects both vegetative production and the activity of organisms. Hot, dry desert regions have sparse vegetation and hence limited organic material available for the soil. The lack of precipitation inhibits chemical weathering leading to coarse textured soil in arid regions. Bacterial activity is limited by the cold temperatures in the tundra causing organic matter to build up.

In the warm and wet tropics, bacterial activity proceeds at a rapid rate, thoroughly decomposing leaf litter. Under the lush tropical forest vegetation, available nutrients are rapidly taken back up by the trees. The high annual precipitation also flushes some organic material from the soil. These factors combine to create soils lacking much organic matter in their upper horizons.

Climate, interacting with vegetation, also affects soil chemistry. Pine forests tend to dominate cool, humid climates. Decomposing pine needles in the presence of water creates a weak acid that strips soluble bases from soil leaving it in an acidic state.

Additionally, pine trees have low nutrient demands so few soil nutrients are taken back up by the trees to be later recycled by decaying needle litter. Broadleaf deciduous trees like oak and maple have higher nutrient demand and thus continually recycle soil nutrients keeping soils high in soluble bases.

Topography

Topography has a significant impact on soil formation as it determines runoff of water, and its orientation affects microclimate which in turn affects vegetation. For soil to form, the parent material needs to lie relatively undisturbed so soil horizon processes can proceed. Water moving across the surface strips parent material away impeding soil development. Water erosion is more effective on steeper, unvegetated slopes.

Effect on Soil Erosion

Slope angle and length affects runoff generated when rain falls to the surface. Examine the diagram below showing the relationship between hill slope position, runoff, and erosion.

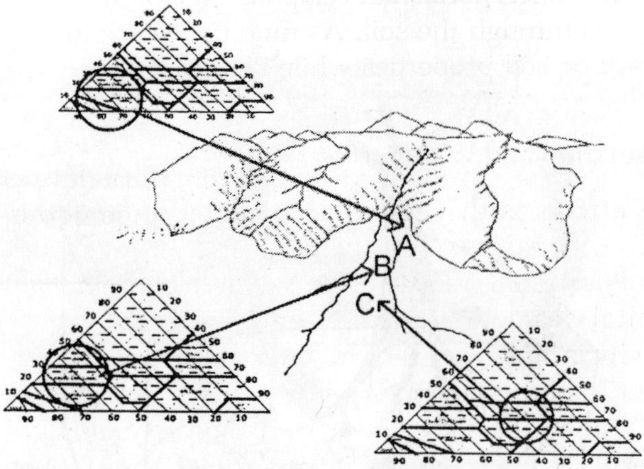

Fig:10. Hill Slope Position, Runoff & Erosion

The amount of water on a particular hill slope segment is dependent on what falls from precipitation and what runs into it from an upslope hill slope segment. As water runs down slope, the water that has accumulated in segment A runs off adding to what falls into segment B by precipitation. The water in B runs into C, and C into D, and so on. The amount of water increases in the down slope direction as water is contributed of water from upslope segments.

The velocity of the water increases as well as it moves towards the base of the slope. As a result, the amount and velocity of water, and hence rate of erosion increases as you near the base of the slope. Rather than infiltrating into the soil to promote weathering and soil development, water runs off.

Erosion causes stripping of soil thus preventing parent material to stay in place to develop into a soil. So we should expect to find weakly developed soil at the mid and near the bottom of the slope.

Effect on Deposition and Soil Texture

Water velocity not only determines the rate of erosion but the deposition of soil material in suspension too. Sites A, B, and C, are located progressively further from the base of a slope. As water empties from a mountain stream, its velocity starts to decrease. The largest size particles, like sand, are the first to drop out of suspension (Site A).

Fine, clay size particles can be carried further away from the base of the slope before they are deposited. As a result, coarse textured soils tend to be found near the base of the mountain and fine textured soils are located further away (Site C).

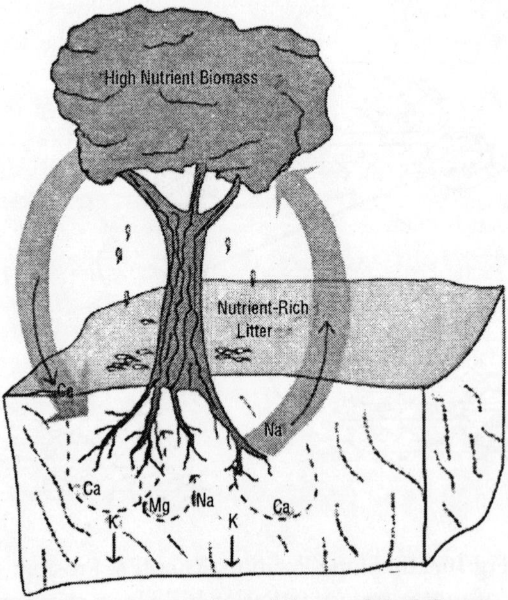

Fig:11. Location, Deposition and Soil Texture

Microclimatic Effects

Hill slope orientation affects the microclimate of a place. As the slope of the surface increases, so does the local sun angle, up to a point. As the local sun angle increases, the intensity of heating increases, causing warmer surface temperatures and, likely, increased evaporation. Orientation of the hill slope is certainly important too. Those slopes which face into the sun receive more insolation than those facing away. Thus inclined surfaces facing into the sun tend to be warmer and drier, than flatter surfaces facing way from the sun. The microclimate also impact vegetation type.

Organisms

Organism, both plant and animal, play an important role in the development and composition of soil. Organisms add organic matter, aid decomposition, weathering and nutrient cycling. The richness and diversity of soil organisms and plant life that grows on the surface is, of course, also tied to climate.

Nutrient Cycling

Biotic elements of the environment need life-sustaining nutrients that find their origin in soil. Upon their death, organisms return these nutrients to soil to be taken up again by other plants and animals. Hence there is a constant cycling of nutrients between organisms and soils. This cycling refreshes and maintains the nutrient status of soils. Without it, soluble nutrients would be leached from soil, decreasing soil's ability to support life.

The degree to which nutrients are cycled depends on the needs of the organism occupying a particular place. For instance, broadleaf, deciduous trees like oak and maple generally have high nutrient demand creating surface litter rich in nutrients when leaves die and fall to the forest floor.

Decomposition of the litter releases the nutrients back into soil for the tree to take back up. Thus soils under these kinds of forests tend to be high in soluble bases and nutrients.

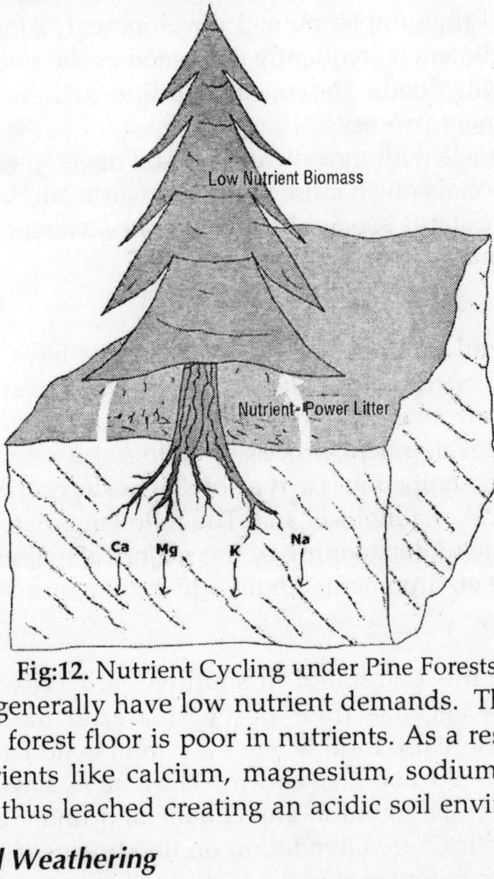

Fig:12. Nutrient Cycling under Pine Forests

Pine trees generally have low nutrient demands. The decaying litter that falls to the forest floor is poor in nutrients. As a result, little cycling of soluble nutrients like calcium, magnesium, sodium, and potassium occurs and are thus leached creating an acidic soil environment.

Organisms and Weathering

Soil organisms also affect weathering. The decomposition of pine needles creates a weak acid that can strip soluble ions from soil. Burrowing animals create passage ways through soil to help aerate and allow water to infiltrate into it. Burrowing animals help translocate materials and fertilise soil at depth.

Time

As time passes, the weathering processes continue to act on soil parent material to break it down and decompose it. Horizon development processes continue to differentiate layers in the soil profile by their physical and chemical properties. As a result, older more mature soils have well-developed sequence of horizons, though some may undergo so much weathering and leaching that visually distinct layers may be hard to see. This is a notable characteristic of oxisols.

Some geological processes keep soils from developing by constantly altering the surface and thus not allowing parent material to weather over a significant period of time. For instance, erosion of hillsides constantly removes material thus impeding soil development. Along the channels of rivers, new sediment is frequently deposited as the river spills out onto its floodplain during floods. The constant addition of new material restarts the soil development process.

Climate interacts with time during the soil development process. Soil development proceeds much more rapidly in warm and wet climates thus reaching a mature status sooner. In cold climates, weathering is impeded and soil development takes much longer.

SOIL ORDERS

The are several different ways in which soils have been classified. Soils are classified on the basis of *diagnostic horizons,* that are different from the O, A, B, C horizons. A diagnostic horizon has a unique feature that reflects the soil development processes acting at a site. For instance, a mollic epipedon (an epipedon is a type of diagnostic horizon) is a organic-rich horizon typical of a mollisol soil. The following material is intended to give you a basic understanding of the major categories of soils called *soil orders* and the environments under which they form.

Entisol

Entisols, like this one found in southwestern Wisconsin, are soils lacking horizons because their parent material has only recently accumulated. Entisols also form where the parent material is quartz sand, in which horizons do not easily form. They have a wide geographic distribution and can be found in any climate and under any vegetation. Entisols and Inceptisols are often found on floodplains, delta deposits, or steep slopes where parent material has difficulty accumulating.

Inceptisol

Inceptisols are soils just starting to show horizon development because the soil is quite young. The differentiation of layers in an inceptisol formed on colluvium in West Virginia on the right. Inceptisols, like

Entisols, are found in any type of environment and are commonly found forming in alluvium on floodplains and delta deposits.

Histosol

Histosols have a very high content of organic matter in the dark upper layer of the profile. Found in many different environments from the tundra to the tropics, Histosols form in places where organic matter is slow to decompose and thus accumulates over time such as bogs and swamps. They are often "mined" for peat which is dried and burned as fuel.

Aridisol

Aridisols are soils of arid and semiarid environments where moisture is scarce. They are typically light in colour as there is little vegetation to add organic matter to the soil profile. A negative moisture balance in these soils inhibits eluviation.

Calcification and salinisation are important soil forming processes acting in these soils. Soil horizons are weakly developed and sodium is often high in concentration making them alkaline.

The coarse texture of aridisols makes it difficult to retain much moisture. Aridisols can be quite fertile soil if irrigation is properly used. Used improperly, a salt crust can form on the soil. Most aridisols are used for grazing.

Andisol

Andisols are soils developing in parent material containing at least fifty per cent volcanic ash. The layers of ash can be seen in this Andisol from Hawaii. Naturally fertile soils, they support a dense natural cover in moist climates. Andisols occur around individual volcanoes created from andesite-rich magma. They are common on the volcanic islands and mountains of "The Ring of Fire", that encircles the Pacific Ocean from North America through Japan.

Vertisol

Vertisols are dark black soils rich in expandable clay minerals. The clay readily swells upon wetting and shrinks when dried. Though found in every type of climate, they are often found in steppe and wet/dry tropical climates where the soil develops deep cracks as it dries. Surface fragments fall into the cracks and are "swallowed" when the soil swells upon wetting. The soil then develops an "inverted profile" with organic material that is typically located near the surface of the profile is now found at depth.

ORGANIC MATTER DECOMPOSITION AND THE SOIL FOOD WEB

Soil Organic Matter

When plant residues are returned to soil, various organic compounds undergo decomposition. Decomposition is a biological process that includes the physical breakdown and biochemical transformation of complex organic molecules of dead material into simpler organic and inorganic molecules. The continual addition of decaying plant residues to soil surface contributes to the biological activity and the carbon cycling process in the soil. Breakdown of soil organic matter and root growth and decay also contribute to these processes. Carbon cycling is the continuous transformation of organic and inorganic carbon compounds by plants and micro- and macro-organisms between the soil, plants and the atmosphere.

Decomposition of organic matter is largely a biological process that occurs naturally. Its speed is determined by three major factors: soil organisms, the physical environment and the quality of the organic matter. In the decomposition process, different products are released: carbon dioxide (CO_2), energy, water, plant nutrients and resynthesised organic carbon compounds. Successive decomposition of dead material and modified organic matter results in the formation of a more complex organic matter called humus. This process is called humification. Humus affects soil properties. As it slowly decomposes, it colours the soil darker; increases soil aggregation and aggregate stability; increases the CEC (the ability to attract and retain nutrients); and contributes N, P and other nutrients.

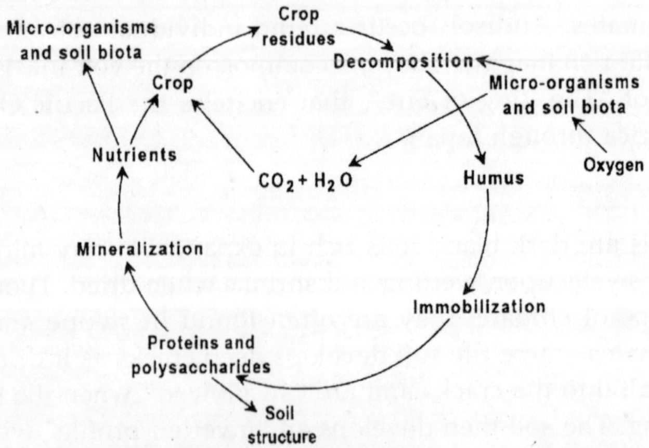

Fig:13. Carbon Cycle

Soil organisms, including microorganisms, use soil organic matter as food. As they break down the organic matter, any excess nutrients (N, P and S) are released into the soil in forms that plants can use. This release process is called mineralisation. The waste products produced by micro-organisms are also soil organic matter. This waste material is less decomposable than the original plant and animal material, but it can be used by a large number of organisms. By breaking down carbon structures and rebuilding new ones or storing the C into their own biomass, soil biota plays the most important role in nutrient cycling processes and, thus, in the ability of a soil to provide the crop with sufficient nutrients to harvest a healthy product. The organic matter content, especially the more stable humus, increases the capacity to store water and store (sequester) C from the atmosphere.

The Soil Food Web

The soil ecosystem can be defined as an interdependent life-support system composed of air, water, minerals, organic matter, and macro- and micro-organisms, all of which function together and interact closely. The organisms and their interactions enhance many soil ecosystem functions and make up the soil food web. The energy needed for all food webs is generated by primary producers: the plants, lichens, moss, photosynthetic bacteria and algae that use sunlight to transform CO_2 from the atmosphere into carbohydrates. Most other organisms depend on the primary producers for their energy and nutrients; they are called consumers. Soil life plays a major role in many natural processes that determine nutrient and water availability for agricultural productivity.

The primary activities of all living organisms are growing and reproducing. By-products from growing roots and plant residues feed soil organisms. In turn, soil organisms support plant health as they decompose organic matter, cycle nutrients, enhance soil structure and control the populations of soil organisms, both beneficial and harmful (pests and pathogens) in terms of crop productivity. The living part of soil organic matter includes a wide variety of micro-organisms such as bacteria, viruses, fungi, protozoa and algae. It also includes plant roots, insects, earthworms, and larger animals such as moles, mice and rabbits that spend part of their life in the soil. The living portion represents about 5 per cent of the total soil organic matter. Microorganisms, earthworms and insects help break down crop residues and manures by ingesting them and mixing them with the minerals in the soil, and in the process recycling energy and plant nutrients.

Sticky substances on the skin of earthworms and those produced by fungi and bacteria help bind particles together. Earthworm casts are also more strongly aggregated (bound together) than the surrounding soil as a result of the mixing of organic matter and soil mineral material, as well as the intestinal mucus of the worm. Thus, the living part of soil is responsible for keeping air and water available, providing plant nutrients, breaking down pollutants and maintaining soil structure.

The composition of soil organisms depends on the food source (which in turn is season dependent). Therefore, the organisms are neither uniformly distributed through the soil nor uniformly present all year. However, in some cases their biogenic structures remain. Each species and group exists where it can find appropriate food supply, space, nutrients and moisture.

Organisms occur wherever organic matter occurs. Therefore, soil organisms are concentrated: around roots, in litter, on humus, on the surface of soil aggregates and in spaces between aggregates. For this reason, they are most prevalent in forested areas and cropping systems that leave a lot of biomass on the surface. The activity of soil organisms follows seasonal as well as daily patterns.

Not all organisms are active at the same time. Most are barely active or even dormant. Availability of food is an important factor that influences the level of activity of soil organisms and thus is related to land use and management.

Practices that increase numbers and activity of soil organisms include: no-tillage or minimal tillage; and the maintenance of plant and annual residues that reduce disturbance of soil organisms and their habitat and provide a food supply. Different groups of organisms can be distinguished in the soil.

DECOMPOSITION PROCESS

Fresh residues consist of recently deceased microorganisms, insects and earthworms, old plant roots, crop residues, and recently added manures. Crop residues contain mainly complex carbon compounds originating from cell walls (cellulose, hemicellulose, etc.). Chains of carbon, with each carbon atom linked to other carbons, form the "backbone" of organic molecules.

These carbon chains, with varying amounts of attached oxygen, H, N, P and S, are the basis for both simple sugars and amino acids and more complicated molecules of long carbon chains or rings. Depending on their chemical structure, decomposition is rapid (sugars, starches and proteins), slow (cellulose, fats, waxes and resins) or very slow (lignin).

During the decomposition process, microorganisms convert the carbon structures of fresh residues into transformed carbon products in soil. There are many different types of organic molecules in soil. Some are simple molecules that have been synthesised directly from plants or other living organisms.

These relatively simple chemicals, such as sugars, amino acids, and cellulose are readily consumed by many organisms. For this reason, they do not remain in soil for a long time. Other chemicals such as resins and waxes also come directly from plants, but are more difficult for soil organisms to break down.

Humus is the result of successive steps in the decomposition of organic matter. Because of the complex structure of humic substances, humus cannot be used by many microorganisms as an energy source and remains in soil for a relatively long time.

Non-Humic Substances: Significance and Function

Non-humic organic molecules are released directly from cells of fresh residues, such as proteins, amino acids, sugars, and starches. This part of soil organic matter is the active, or easily decomposed, fraction. This active fraction is influenced strongly by weather conditions, moisture status of soil, growth stage of the vegetation, addition of organic residues, and cultural practices, such as tillage.

It is the main food supply for various organisms in soil. Carbohydrates occur in soil in three main forms: free sugars in the soil solution, cellulose and hemicellulose; complex polysaccharides; and polymeric molecules of various sizes and shapes that are attached strongly to clay colloids and humic substances.

The simple sugars, cellulose and hemicellulose, may constitute 5–25 per cent of the organic matter in most soils, but are easily broken down by microorganisms. Polysaccharides (repeating units of sugar-type molecules connected in longer chains) promote better soil structure through their ability to bind inorganic soil particles into stable aggregates. Research indicates that the heavier polysaccharide molecules may be more important in promoting aggregate stability and water infiltration than the lighter molecules.

Some sugars may stimulate seed germination and root elongation. Other soil properties affected by polysaccharides include CEC, anion retention and biological activity. Soil lipids form a very diverse group of materials, of which fats, waxes and resins make up 2–6 per cent of soil organic matter.

The significance of lipids arises from the ability of some compounds to act as growth hormones. Others may have a depressing effect on plant growth. Soil N occurs mainly (> 90 per cent) in organic forms as amino acids, nucleic acids and amino sugars. Small amounts exist in the form of amines, vitamins, pesticides and their degradation products, etc. The rest is present as ammonium (NH_4^-) and is held by the clay minerals.

Compounds and Function of Humus

Humus or humified organic matter is the remaining part of organic matter that has been used and transformed by many different soil organisms. It is a relatively stable component formed by humic substances, including humic acids, fulvic acids, hymatomelanic acids and humins. It is probably the most widely distributed organic carbon-containing material in terrestrial and aquatic environments.

Humus cannot be decomposed readily because of its intimate interactions with soil mineral phases and is chemically too complex to be used by most organisms. One of the most striking characteristics of humic substances is their ability to interact with metal ions, oxides, hydroxides, mineral and organic compounds, including toxic pollutants, to form water-soluble and water-insoluble complexes.

Through the formation of these complexes, humic substances can dissolve, mobilise and transport metals and organics in soils and waters, or accumulate in certain soil horizons.

This influences nutrient availability, especially those nutrients present at microconcentrations only. Accumulation of such complexes can contribute to a reduction of toxicity, e.g., of aluminium (Al) in acid soils or the capture of pollutants – herbicides such as Atrazine or pesticides such as Tefluthrin – in the cavities of the humic substances.

Humic and fulvic substances enhance plant growth directly through physiological and nutritional effects. Some of these substances function as natural plant hormones (auxines and gibberillins) and are capable of improving seed germination, root initiation, uptake of plant nutrients and can serve as sources of N, P and S. Indirectly, they may affect plant growth through modifications of physical, chemical and biological properties of soil, for example, enhanced soil water holding capacity and CEC, and improved tilth and aeration through good soil structure.

About 35–55 per cent of the non-living part of organic matter is humus. It is an important buffer, reducing fluctuations in soil acidity and nutrient availability. Compared with simple organic molecules, humic substances are very complex and large, with high molecular weights. The characteristics of the well-decomposed part of the organic matter, the humus, are very different from those of simple organic molecules. While much is known about their general chemical composition, the relative significance of the various types of humic materials to plant growth is yet to be established.

Humus consists of different humic substances:

- *Fulvic acids:* The fraction of humus that is soluble in water under all pH conditions. Their colour is commonly light yellow to yellow-brown.
- *Humic acids:* The fraction of humus that is soluble in water, except for conditions more acid than pH 2. Common colours are dark brown to black.

- *Humin:* The fraction of humus that is not soluble in water at any pH and that cannot be extracted with a strong base, such as sodium hydroxide (NaOH). Commonly black in colour.

The term acid is used to describe humic materials because humus behaves like weak acids. Fulvic and humic acids are complex mixtures of large molecules.

Humic acids are larger than fulvic acids. Research suggests that the different substances are differentiated from each other on the basis of their water solubility.

Fulvic acids are produced in the earlier stages of humus formation. The relative amounts of humic and fulvic acids in soils vary with soil type and management practices. The humus of forest soils is characterised by a high content of fulvic acids, while the humus of agricultural and grassland areas contains more humic acids.

Chapter 5

Soil and Water

SOIL

Soil Composition

When dry soil is crushed in the hand, it can be seen that it is composed of all kinds of particles of different sizes. Most of these particles originate from the degradation of rocks; they are called mineral particles. Some originate from residues of plants or animals (rotting leaves, pieces of bone, etc.), these are called organic particles (or organic matter).

The soil particles seem to touch each other, but in reality have spaces in between. These spaces are called pores. When the soil is "dry", the pores are mainly filled with air. After irrigation or rainfall, the pores are mainly filled with water. Living material is found in soil. It can be live roots as well as beetles, worms, larvae, etc. They help to aerate soil and thus create favourable growing conditions for the plant roots.

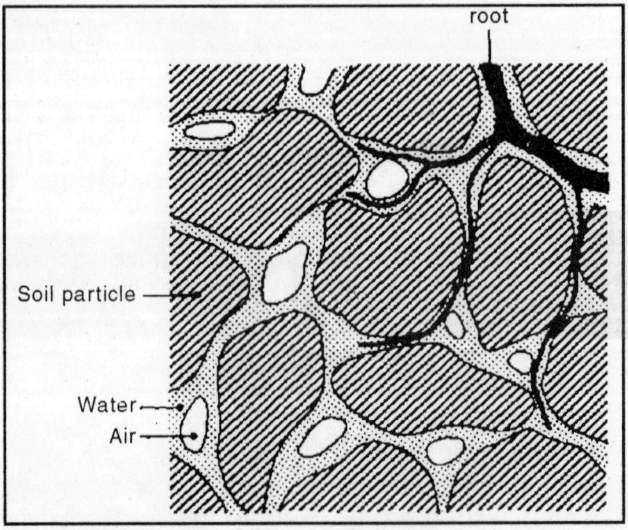

Fig:1. The Composition of Soil

Soil Profile

If a pit is dug in the soil, at least 1 m. deep, various layers, different in colour and composition can be seen. These layers are called horizons. This succession of horizons is called the profile of soil.

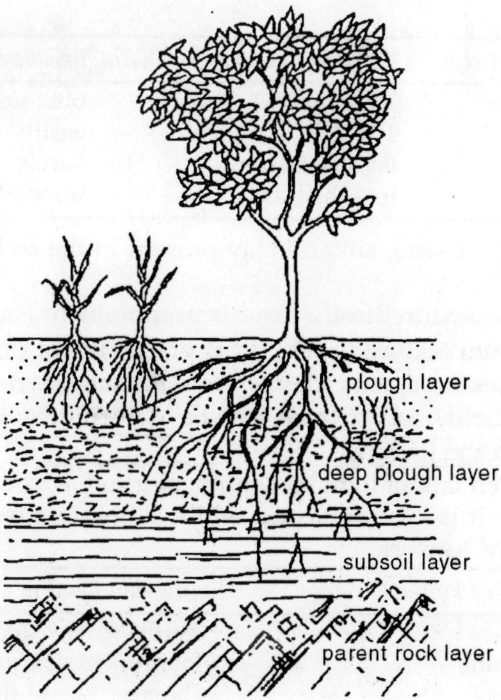

Fig:2. The Soil Profile

A very general and simplified soil profile can be described as follows:

- *The plough layer* (20 to 30 cm. thick): is rich in organic matter and contains many live roots. This layer is subject to land preparation (e.g., ploughing, harrowing, etc.) and often has a dark colour (brown to black).
- *The deep plough layer:* contains much less organic matter and live roots. This layer is hardly affected by normal land preparation activities. The colour is lighter, often grey and sometimes mottled with yellowish or reddish spots.
- *The subsoil layer:* hardly any organic matter or live roots are to be found. This layer is not very important for plant growth as only a few roots will reach it.

- *The parent rock layer:* consists of rock, from the degradation of which the soil was formed. This rock is sometimes called parent material. The depth of the different layers varies widely: some layers may be missing altogether.

Soil Texture

The mineral particles of soil differ widely in size and can be classified as follows:

Name of the Particles	*Size Limits in mm*	*Distinguisable with Naked Eye*
gravel	larger than 1	obviously
sand	1 to 0.5	easily
silt	0.5 to 0.002	barely
clay	less than 0.002	impossible

The amount of sand, silt and clay present in the soil determines the soil texture.

- In coarse textured soils: sand is predominant (sandy soils).
- In medium textured soils: silt is predominant (loamy soils).
- In fine textured soils: clay is predominant (clayey soils).
- In the field, soil texture can be determined by rubbing soil between the fingers.

Farmers often talk of light soil and heavy soil. A coarse-textured soil is light because it is easy to work, while a fine-textured soil is heavy because it is hard to work.

Expression Used by the Farmer	*Expression Used in Literature*	
light	sandy	coarse
medium	loamy	medium
heavy	clayey	fine

The texture of a soil is permanent, the farmer is unable to modify or change it.

Soil Structure

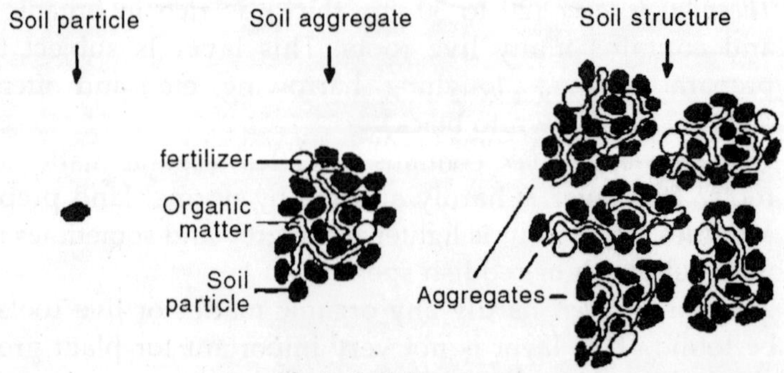

Fig:3. Soil Structure

Soil structure refers to the grouping of soil particles (sand, silt, clay, organic matter and fertilizers) into porous compounds. These are called aggregates. Soil structure also refers to the arrangement of these aggregates separated by pores and cracks. The basic types of aggregate arrangements, granular, blocky, prismatic and massive structure.

When present in the topsoil, a massive structure blocks the entrance of water; seed germination is difficult due to poor aeration. On the other hand, if the topsoil is granular, the water enters easily and the seed germination is better.

In a prismatic structure, movement of the water in soil is predominantly vertical and therefore the supply of water to the plant roots is usually poor.

Unlike texture, soil structure is not permanent. By means of cultivation practices (ploughing, ridging, etc.), the farmer tries to obtain a granular topsoil structure for his fields.

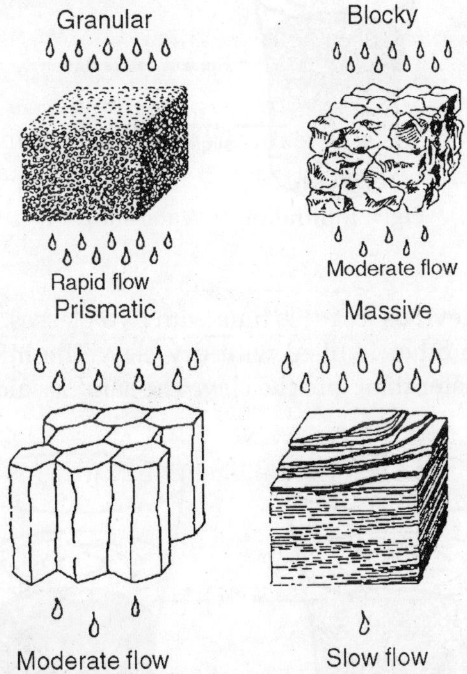

Fig:4. Some Examples of Soil Structures

ENTRY OF WATER INTO SOIL

Infiltration Process

When rain or irrigation water is supplied to a field, it seeps into soil. This process is called infiltration.

Infiltration can be visualised by pouring water into a glass filled with dry powdered soil, slightly tamped. The water seeps into soil; the colour of soil becomes darker as it is wetted.

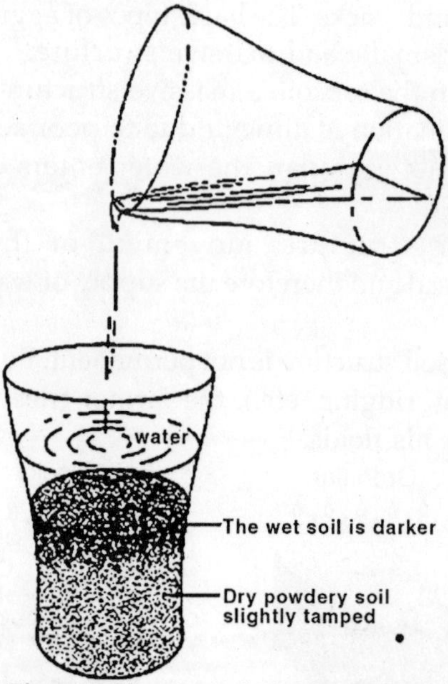

Fig:5. Infiltration of Water into Soil

Infiltration Rate

Repeat the previous test, this time with two glasses. One is filled with dry sand and the other is filled with dry clay. The infiltration of water into the sand is faster than into the clay. The sand is said to have a higher infiltration rate.

Fig:6. The Same Amount of Water is Supplied to Each Glass

The infiltration rate of a soil is the velocity at which water can seep into it. It is commonly measured by the depth (in mm) of the water layer that the soil can absorb in an hour. An infiltration rate of 15 mm/hour means that a water layer of 15 mm on the surface of the soil, will take one hour to infiltrate.

Fig:7. After One Hour the Water has Infiltrated in the Sand

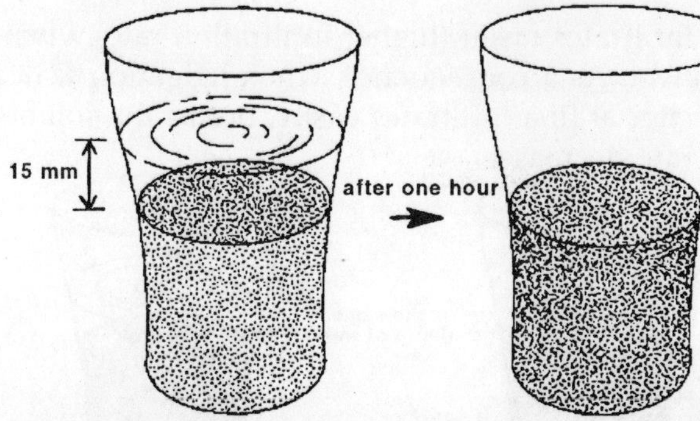

Fig:8. Soil with an Infiltration Rate of 15 mm/hour

A range of values for infiltration rates is given below:

Low infiltration rate	Less than 15 mm/hour
Medium infiltration rate	15 to 50 mm/hour
High infiltration rate	More than 50 mm/hour

Factors Influencing the Infiltration Rate

The infiltration rate of a soil depends on factors that are constant, such as the soil texture. It also depends on factors that vary, such as the soil moisture content.

Soil Texture

Coarse textured soils have mainly large particles in between which there are large pores. On the other hand, fine textured soils have mainly small particles in between which there are small pores. In coarse soils, the rain or irrigation water enters and moves more easily into larger pores; it takes less time for the water to infiltrate into soil. In other words, infiltration rate is higher for coarse textured soils than for fine textured soils.

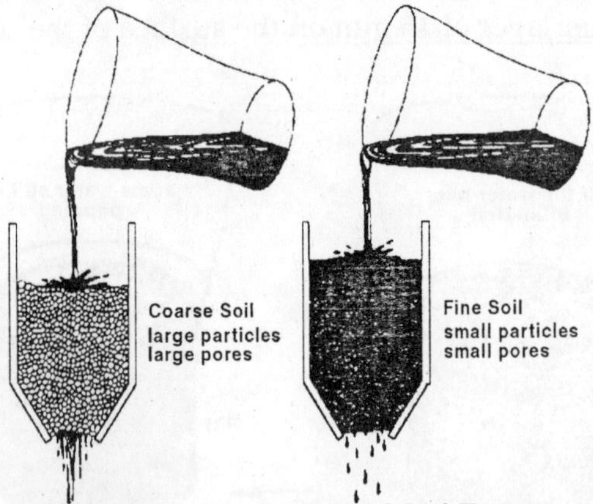

Fig:9. Infiltration Rate and Soil Texture

The Soil Moisture Content

The water infiltrates faster (higher infiltration rate) when soil is dry, than when it is wet. As a consequence, when irrigation water is applied to a field, the water at first infiltrates easily, but as the soil becomes wet, the infiltration rate decreases.

Fig:10. Infiltration Rate and Soil Moisture Content

Soil Structure

Generally speaking, water infiltrates quickly (high infiltration rate) into granular soils but very slowly (low infiltration rate) into massive and compact soils. Because the farmer can influence soil structure (by means of cultural practices), he can also change the infiltration rate of his soil.

SOIL MOISTURE CONDITIONS

Soil Moisture Content

The soil moisture content indicates the amount of water present in soil. It is commonly expressed as the amount of water (in mm of water depth) present in a depth of one metre of soil. For example: when an amount of water (in mm of water depth) of 150 mm is present in a depth of one metre of soil, the soil moisture content is 150 mm/m.

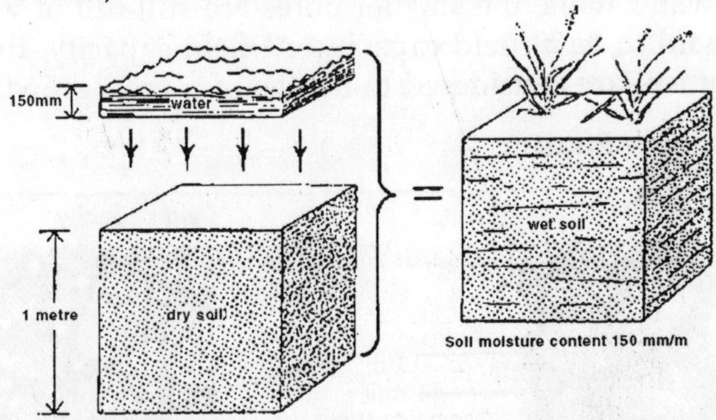

Fig:11. A Soil Moisture Content of 150 mm/m

The soil moisture content can also be expressed in per cent of volume. In the example above, 1 m^3 of soil (e.g., with a depth of 1 m and a surface area of 1 m^2) contains 0.150 m^3 of water (e.g., with a depth of 150 mm = 0.150 m and a surface area of 1 m^2). This results in a soil moisture content

in volume per cent of: $\frac{0.150m^3}{1m^3} \times 100\% = 15\%$

Thus, a moisture content of 100 mm/m corresponds to a moisture content of 10 volume per cent.

Note: The amount of water stored in the soil is not constant with time, but may vary.

Saturation

During a rain shower or irrigation application, soil pores will fill with water. If all soil pores are filled with water the soil is said to be saturated. There is no air left in soil. It is easy to determine in the field if a soil is saturated. If a handful of saturated soil is squeezed, some (muddy) water will run between the fingers. Plants need air and water in soil. At saturation, no air is present and the plant will suffer. Many crops cannot withstand saturated soil conditions for a period of more than 2-5 days. Rice is one of the exceptions to this rule.

The period of saturation of the topsoil usually does not last long. After the rain or the irrigation has stopped, part of the water present in the larger pores will move downward.

This process is called drainage or percolation. The water drained from the pores is replaced by air. In coarse textured (sandy) soils, drainage is completed within a period of a few hours. In fine textured (clayey) soils, drainage may take some (2-3) days.

Field Capacity

After the drainage has stopped, the large soil pores are filled with both air and water while the smaller pores are still full of water. At this stage, soil is said to be at field capacity. At field capacity, the water and air contents of soil are considered to be ideal for crop growth.

Permanent Wilting Point

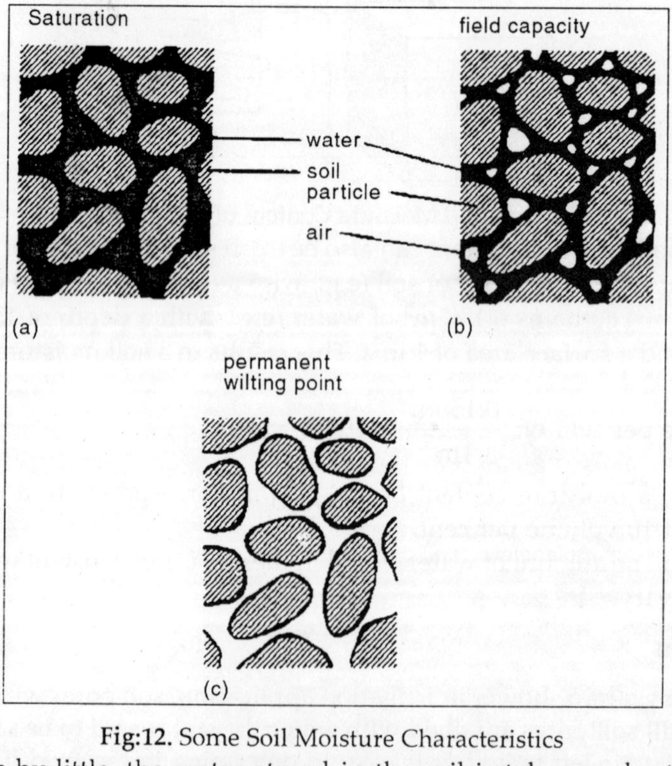

Fig:12. Some Soil Moisture Characteristics

Little by little, the water stored in the soil is taken up by the plant roots or evaporated from the topsoil into the atmosphere. If no additional water is supplied to soil, it gradually dries out. The dryer the soil becomes, the more tightly the remaining water is retained and the more difficult it is for the plant roots to extract it.

At a certain stage, the uptake of water is not sufficient to meet the plant's needs. The plant looses freshness and wilts; the leaves change colour from green to yellow. Finally the plant dies. The soil water content at the stage where the plant dies, is called permanent wilting point. Soil still contains some water, but it is too difficult for the roots to suck it from soil.

AVAILABLE WATER CONTENT

Soil can be compared to a water reservoir for the plants. When soil is saturated, the reservoir is full. However, some water drains rapidly below the rootzone before the plant can use it.

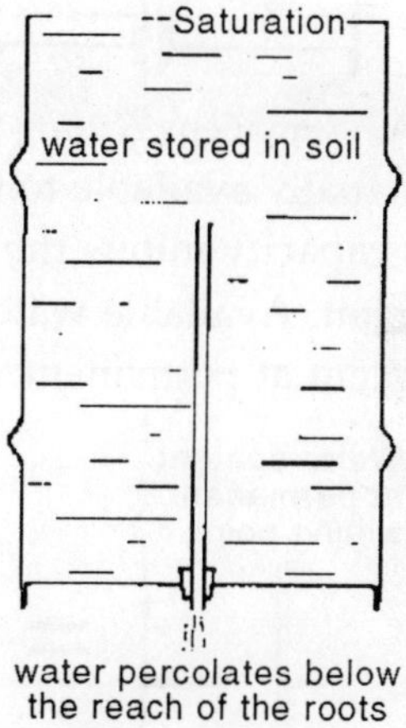

Fig. Saturation

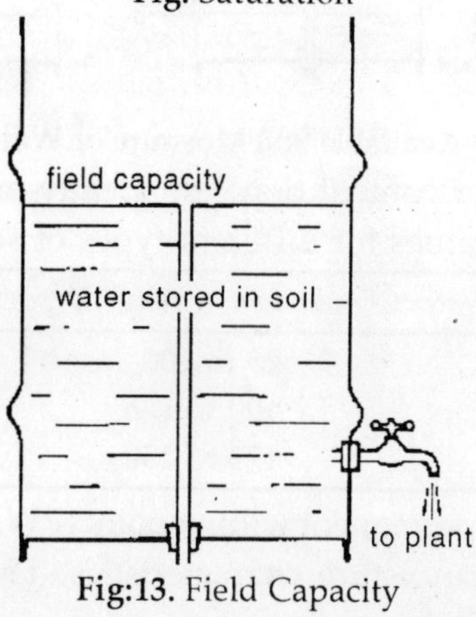

Fig:13. Field Capacity

When this water has drained away, soil is at field capacity. The plant roots draw water from what remains in the reservoir. When soil reaches permanent wilting point, the remaining water is no longer available to the plant.

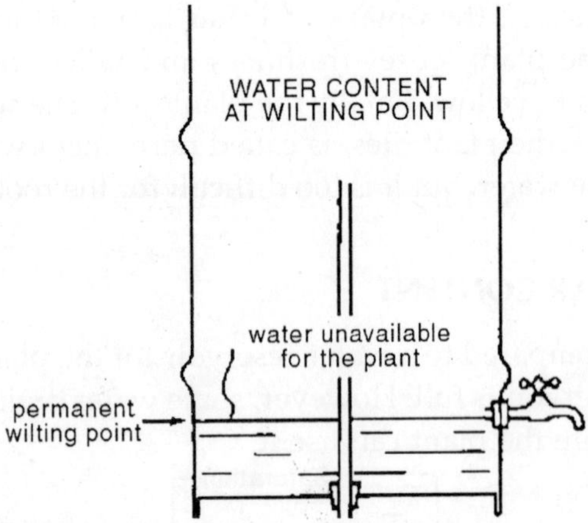

Fig:14A. Permanent Wilting Point

The amount of water actually available to the plant is the amount of water stored in soil at field capacity minus the water that will remain in soil at permanent wilting point. Available water content = water content at field capacity - water content at permanent wilting point.

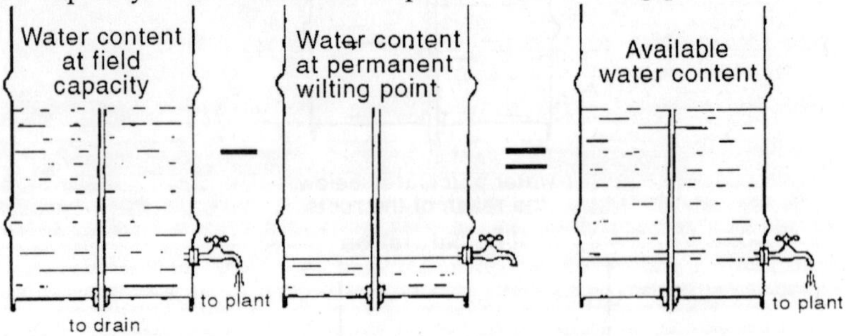

Fig:14B. The Available Soil Moisture or Water Content

The available water content depends greatly on the soil texture and structure. A range of values for different types of soil are:

Soil	*Available Water Content in mm Water Depth per m Soil Depth (mm/m)*
Sand	25 to 100
Loam	100 to 175
Clay	175 to 250

The field capacity, permanent wilting point (PWP) and available water content are called soil moisture characteristics. They are constant for a given soil, but vary widely from one type of soil to another.

GROUNDWATER TABLE

Part of the water applied to the soil surface drains below the rootzone and feeds deeper soil layers which are permanently saturated; the top of the saturated layer is called groundwater table or sometimes just water table.

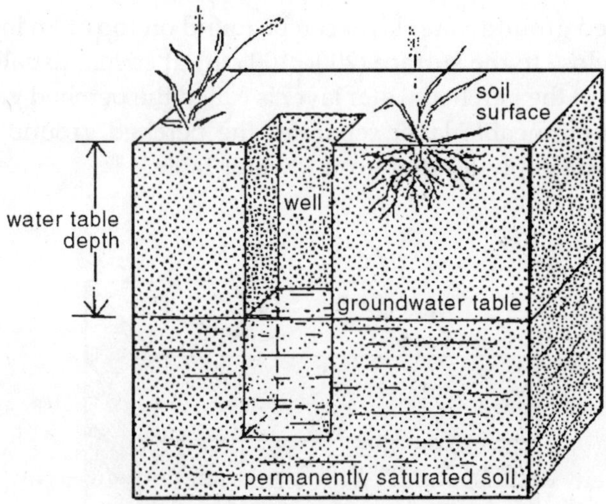

Fig:15. The Groundwater Table

Depth of the Groundwater Table

The depth of the groundwater table varies greatly from place to place, mainly due to changes in topography of the area.

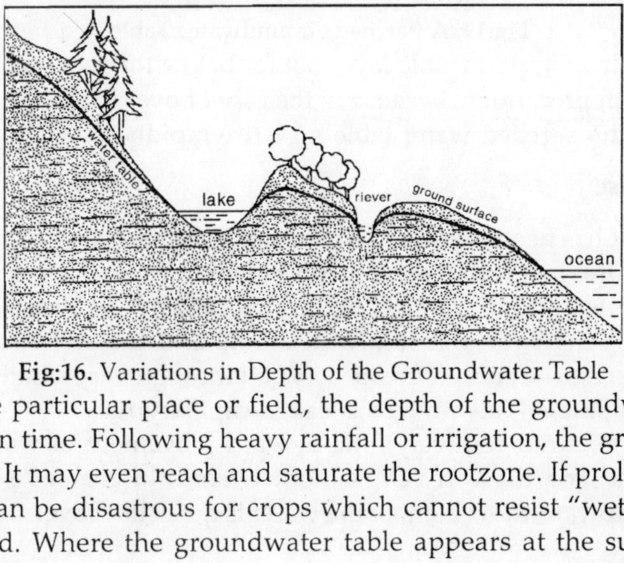

Fig:16. Variations in Depth of the Groundwater Table

In one particular place or field, the depth of the groundwater table may vary in time. Following heavy rainfall or irrigation, the groundwater table rises. It may even reach and saturate the rootzone. If prolonged, this situation can be disastrous for crops which cannot resist "wet feet" for a long period. Where the groundwater table appears at the surface, it is called an open groundwater table. This is the case in swampy areas. The groundwater table can also be very deep and distant from the rootzone, for example, following a prolonged dry period. To keep the rootzone moist, irrigation is then necessary.

Perched Groundwater Table

A perched groundwater layer can be found on top of an impermeable layer rather close to the surface (20 to 100 cm). It covers usually a limited area. The top of the perched water layer is called the perched groundwater table. The impermeable layer separates the perched groundwater layer from the more deeply located groundwater table.

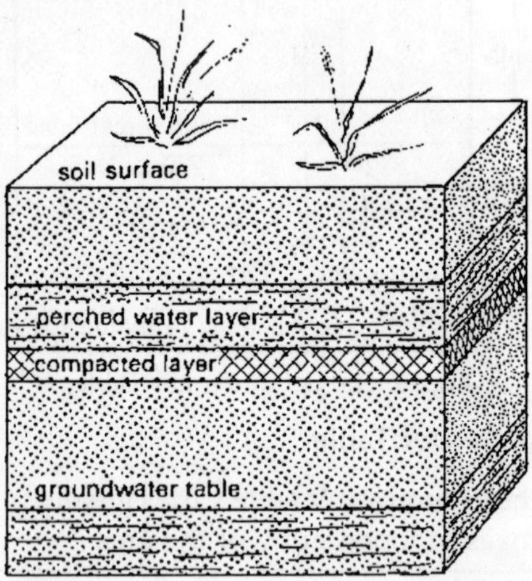

Fig:17. A Perched Groundwater Table

Soil with an impermeable layer not far below the rootzone should be irrigated with precaution, because in the case of over irrigation (too much irrigation), the perched water table may rise rapidly.

Capillary Rise

So far, it has been explained that water can move downward, as well as horizontally. In addition, water can move upward. If a piece of tissue is dipped in water, the water is sucked upward by the tissue. The same process happens with a groundwater table and the soil above it. The groundwater can be sucked upward by soil through very small pores that are called capillars. This process is called capillary rise.

In fine textured soil (clay), the upward movement of water is slow but covers a long distance. On the other hand, in coarse textured soil (sand), the upward movement of the water is quick but covers only a short distance.

Soil Texture	*Capillary Rise (in cm)*
Coarse (sand)	20 to 50 cm
Medium	50 to 80 cm
Fine (clay)	More than 80 cm up to several metres

SOIL EROSION BY WATER

Erosion is the transport of soil from one place to another. Climatic factors such as wind and rain can cause erosion, but also under irrigation it may occur.

Over a short period, the process of erosion is almost invisible. However, it can be continuous and the whole fertile top layer of a field may disappear within a few years.

Soil erosion by water depends on:

- *Slope:* steep, sloping fields are more exposed to erosion.
- *Soil structure:* light soils are more sensitive to erosion.
- *The volume or rate of flow of surface runoff water:* larger or rapid flows induce more erosion.

Erosion is usually the heaviest during the early part of irrigation, especially when irrigating on slopes. The dry surface soil, sometimes loosened by cultivation, is easily removed by flowing water. After the first irrigation, soil is moist and settles down, so erosion is reduced. Newly irrigated areas are more sensitive to erosion, especially in their early stages. There are two main types of erosion caused by water: sheet erosion and gully erosion. They are often combined.

Sheet Erosion

Sheet erosion is the even removal of a very thin layer or "sheet" of topsoil from sloping land. It occurs over large areas of land and causes most of the soil losses.

The signs of sheet erosion are:

- Only a thin layer of topsoil; or the subsoil is partly exposed; sometimes even parent rock is exposed.
- Quite large amounts of coarse sand, gravel and pebbles in the arable layer, the finer material has been removed.
- Exposure of the roots.
- Deposit of eroded material at the foot of the slope.

Gully Erosion

Gully erosion is defined as the removal of soil by a concentrated water flow, large enough to form channels or gullies.

These gullies carry water during heavy rain or irrigation and gradually become wider and deeper.

The signs of gully erosion on an irrigated field are:

- Irregular changes in the shape and length of the furrows;
- Accumulation of eroded material at the bottom of the furrows;
- Exposure of plant roots.

ELEMENTS OF TOPOGRAPHY

Slopes

A slope is the rise or fall of the land surface. It is important for the farmer or irrigator to identify the slopes on the land. A slope is easy to recognise in a hilly area. Start climbing from the foot of a hill toward the top, this is called a rising slope. Go downhill, this is a falling slope.

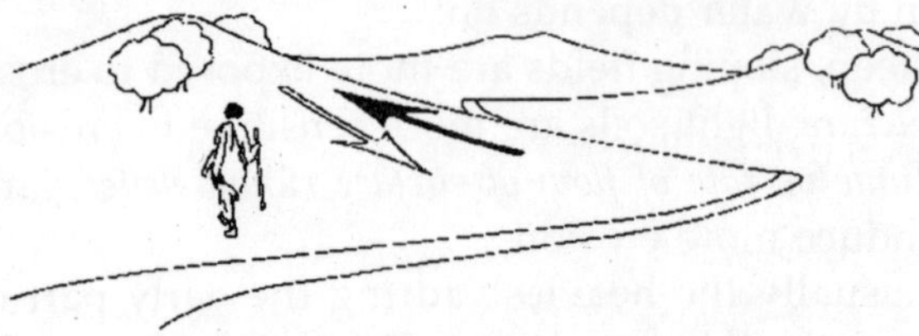

Fig:18. A Rising and a Falling Slope

Flat areas are never strictly horizontal; there are gentle slopes in a seemingly flat area, but they are often hardly noticeable to the naked eye. An accurate survey of the land is necessary to identify these so called "flat slopes".

Method of Expressing Slopes

Slope of a field is expressed as a ratio. It is the vertical distance, or difference in height, between two points in a field, divided by the horizontal distance between these two points. The formula is:

$$\text{Slope} = \frac{\text{height difference (metres)}}{\text{horizontal distance (metres)}}$$

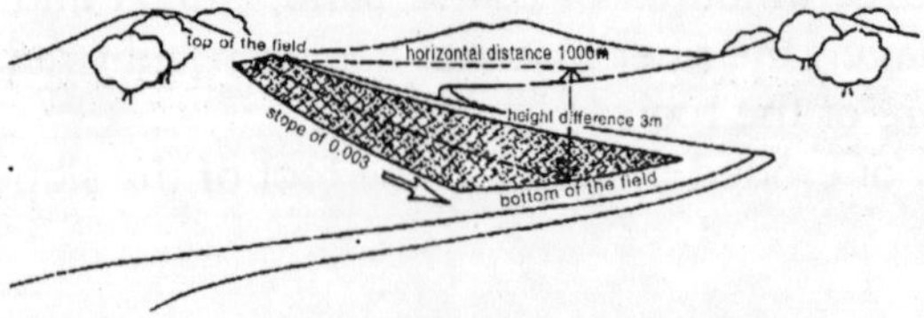

Fig:19. The Dimensions of a Slope

$$\text{Slope} = \frac{\text{height difference(m) between A and B}}{\text{horizontal distance(m) betweeen A and B}}$$

$$= \frac{3\text{m}}{1000\text{m}} = 0.003$$

Slope can also be expressed in per cent; the formula used is then:

$$\text{Slope in per cent} = \frac{\text{height difference (metres)}}{\text{horizontal distance (metres)}} \times 100$$

Using the same measurements:

$$\text{Slope in per cent} = \frac{3\text{m}}{1000\text{m}} \times 100 = 0.3\%$$

Finally, slope can be expressed in per mil; the formula used is then:

$$\text{Slope in per cent} = \frac{\text{height difference (metres)}}{\text{horizontal distance (metres)}} \times 1000...$$

with the figures from the same example:

$$\text{Slope in per cent} = \frac{3\text{m}}{1000\text{m}} \times 1000 = 0.3\%$$

Note: Slope in ‰ = slope in per cent × 10

Question: What is the slope in per cent and in per mil of a field with a horizontal length of 200 m and a height difference of 1.5 m between the top and the bottom?

Answer:

$$\text{Field slope in per cent} = \frac{\text{height difference (metres)}}{\text{horizontal distance (metres)}} \times 100$$

$$= \frac{1.5}{200} \times 100 = 0.75\%$$

Field slope in % = field slope in per cent × 10 = 0.75 × 10 = 7.5%

Question: What is the difference in height between the top and the bottom of a field when the horizontal length of the field is 300 m and the slope is 2%.

Answer:

$$\text{Field slope} = 2\% = 0.002 = \frac{\text{height difference (m)}}{\text{horizontaldistance (m)}}$$

$$= \frac{\text{height difference(m)}}{300\text{m}}$$

Thus: height difference (m) = 0.002 × 300 m = 0.6 m.

Slope	*per cent*	%
Horizontal	0 - 0.2	0 - 2
Very flat	0.2 - 0.5	2 - 5
Flat	0.5 - 1	5 - 10
Moderate	1 - 2.5	10 - 25
Steep	more than 2.5	more than 25

Cross Slopes

Place a book on a table and lift one side of it 4 centimetres from the table. Now, tilt the book sideways (6 cm) so that only one corner of it touches the table.

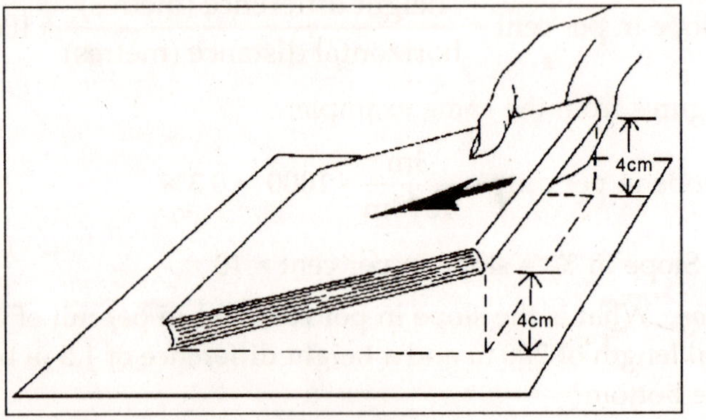

Fig:20. Main Slope

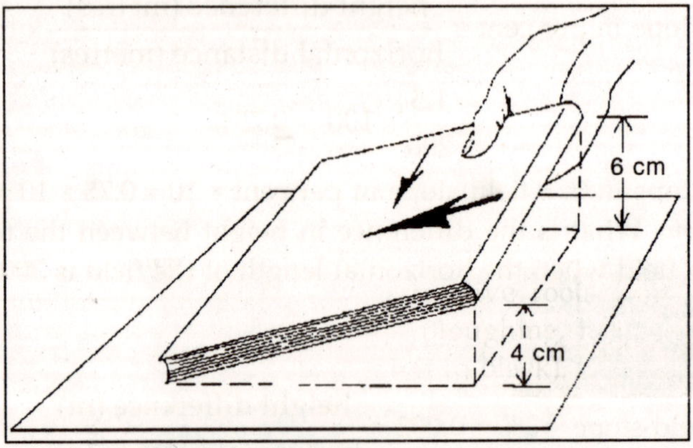

Fig:21. Main Slope and Cross Slope

The thick arrow indicates the direction of what can be called the main slope; the thin arrow indicates the direction of the cross slope, the latter crosses the direction of the main slope. An illustration of the main slope and the cross slope of an irrigated field.

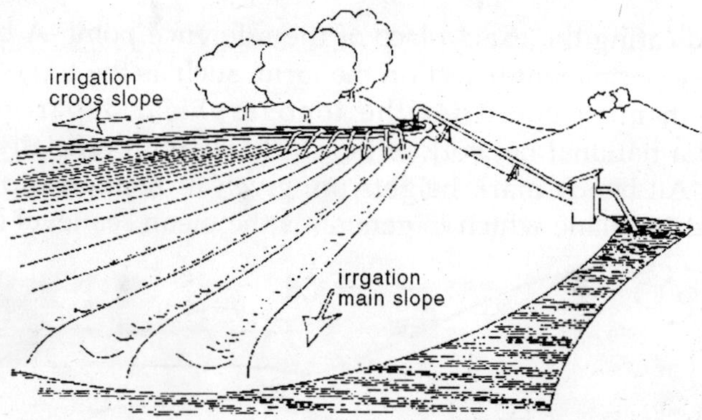

Fig:22. Main Slope and Cross Slope of an Irrigated Field

Elevation of a Point

Point A is at the top of a concrete bridge. Any other point in the surrounding area is higher or lower than A and the vertical distance between the two can be determined.

For example, B is higher than A and the vertical distance between A and B is 2 m. Point C, is lower than A and the vertical distance between A and C is 1 m.

If point A is chosen as a reference point or datum, the elevation of any other point in the field can be defined as the vertical distance between this point and A.

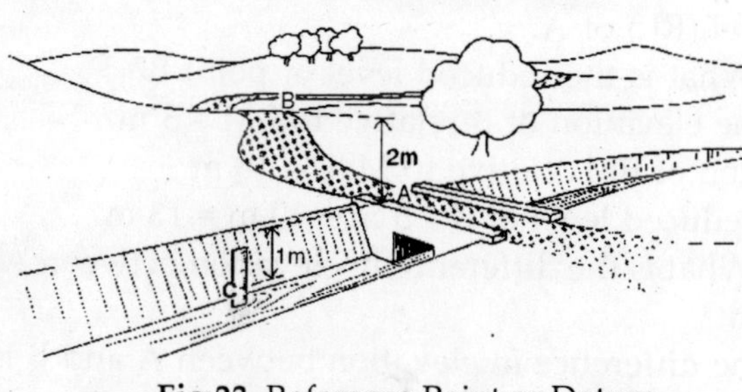

Fig:23. Reference Point or Datum

Thus, the height or elevation of B, in relation to the datum A, is 2 m. and the elevation of C, also related to the datum A, is 1 m. As a reminder that a point is above or below the datum, its elevation is prefixed by the sign + (plus) if it is above the datum, or - (minus) if it is below the datum. Therefore, in relation to the datum A, the elevation of B is +2 m. and the elevation of C is –1 m.

Bench Mark and Mean Sea Level

A bench mark is a permanent mark established in a field to use as a reference point. A bench mark can be a concrete base in which an iron bar is fixed, indicating the exact place of the reference point. A bench mark can also be a permanent object on the farm, such as the top of a concrete structure. In most countries the topographical departments have established a national network of bench marks with officially registered elevations. All bench mark heights are given in relationship to the one national datum plane which in general is the mean sea level (MSL).

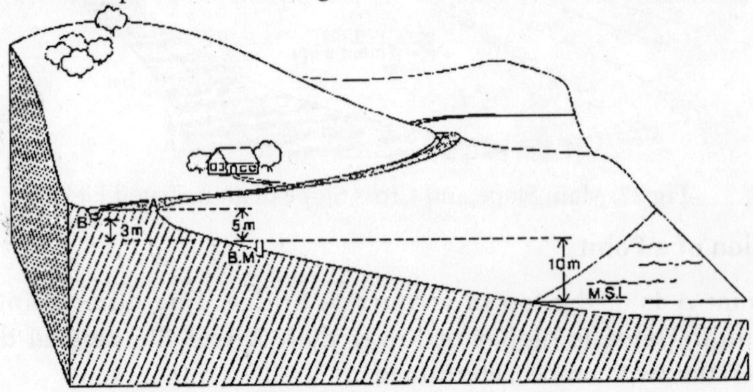

Fig:24. A Bench Mark (BM) and Mean Sea Level (MSL)

Example

The elevation of point A in relation to the bench mark (BM) is 5 metres. The BM elevation relative to the mean sea level (MSL) is 10 m. Thus, the elevation of point A relative to the MSL is 5 m + 10 m = 15 m and is called the reduced level (RL) of A.

Question: What is the reduced level of point B?

Answer: The elevation of B relative to BM = 3 m.

The elevation of BM relative to MSL = 10 m

Thus, the reduced level of B = 3 m + 10 m = 13 m.

Question: What is the difference in elevation between A and B? What does it represent?

Answer: The difference in elevation between A and B is the reduced level of A minus the reduced level of B = 15 m. – 13 m. = 2 m, which represents the vertical distance between A and B.

Contour Lines

A contour line is the imaginary horizontal line that connects all points in a field which have the same elevation. A contour line is imaginary but can be visualised by taking the example of a lake.

The water level of a lake may move up and down, but the water surface always remains horizontal. The level of the water on the shore line of the lake makes a contour line because it reaches points which are all at the same elevation.

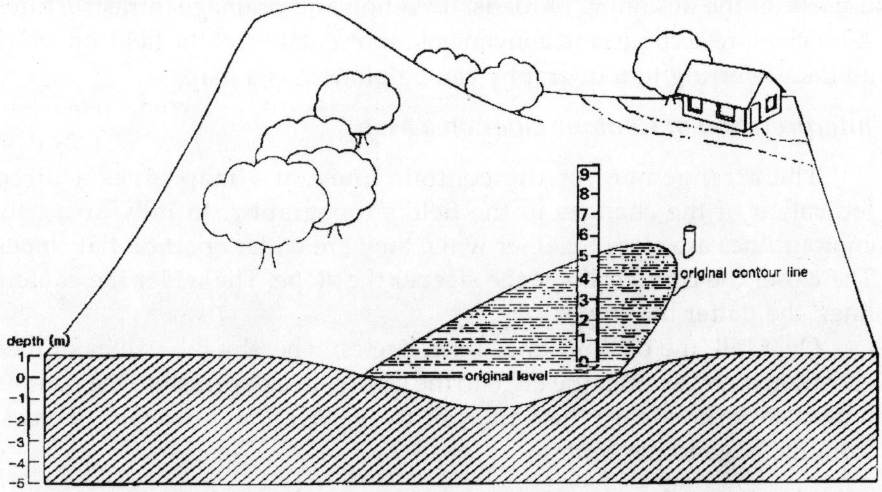

Fig:25. The Shore Line of the Lake forms a Contour Line

Suppose the water level of the lake rises 50 cm. above its original level. The contour line, formed by the shore line, changes and takes a new shape, now joining all the points 50 cm. higher than the original lake level. Contour lines are useful means to the topography of a field on a flat map; the height of each contour line is indicated on the map so that the hills or depressions can be identified.

Maps

Description of a Map

A three-dimensional view of a field with its hills, valleys and depressions; the contour lines have also been indicated.

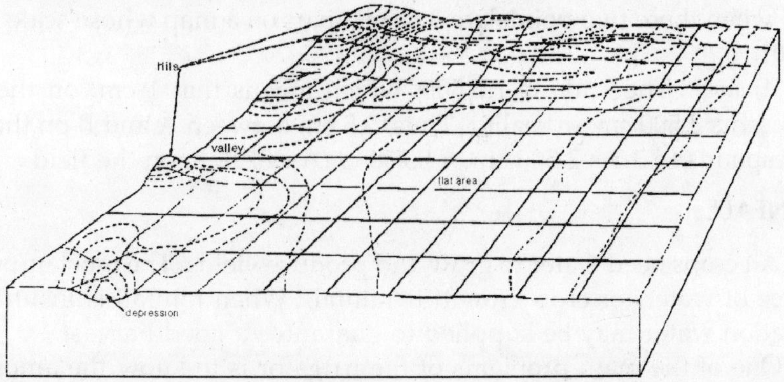

Fig:26. A Three-dimensional View

Such a representation gives a very good idea of what the field looks like in reality. Unfortunately, it requires much skill to draw and is almost useless for the designing of roads, irrigation and drainage infrastructures. A much more accurate and convenient representation of the field, on which all data referring to topography can be plotted, is a map.

Interpretation of Contour Lines on a Map

The arrangement of the contour lines on a map gives a direct indication of the changes in the field's topography. In hilly areas, the contour lines are close together while they are wider apart on flat slopes. The closer the contour lines, the steeper the slope. The wider the contour lines, the flatter the slopes.

On a hill, the contour lines form circles; whereby the values of their elevation increase from the edge to the centre. In a depression, the contour lines also form circles; the values of their elevation, however, decrease from the edge to the centre.

Mistakes in the Contour Lines

Contour lines of different heights can never cross each other. Crossing countour lines would mean that the intersection point has two different elevations, which is impossible. A contour line is continuous; there can never be an isolated piece of contour line somewhere on the map.

Scale of a Map

To be complete and really useful, a map must have a defined scale. The scale is the ratio of the distance between two points on a map and their real distance on the field. A scale of 1 in 5,000 (1:5000) means that 1 cm. measured on the map corresponds to 5,000 cm. (or converted into metres, 50 m.) on the field.

Question: What is the real distance between points A and B on the field when these two points are 3.5 cm. apart on a map whose scale is 1 to 2,500?

Answer: The scale is 1:2,500, which means that 1 cm. on the map represents 2,500 cm. in reality. Thus, 3.5 cm between A and B on the map corresponds to 3.5 × 2,500 cm. = 8,750 cm. or 87.5 m. on the field.

RAINFALL

All crops need water to grow and produce yields. The most important source of water for crop growth is rainfall. When rainfall is insufficient, irrigation water may be supplied to guarantee a good harvest.

One of the main problems of the irrigator is to know the amount of water that has to be applied to the field to meet the water needs of the crops; in other words the irrigation requirement needs to be determined. Too much water means a waste of water which is so precious in arid countries. It can also lead to a rise of the groundwater table and an undesirable saturation of the rootzone. Too little water during the growing season causes the plants to wilt.

Long periods during which the water supply is insufficient, result in loss of yield or even crop failure. In addition, the irrigation requirement needs to be determined for proper design of the irrigation system and for establishment of the irrigation schedules.

The primary source of water for agricultural production, for large parts of the world, is rainfall or precipitation. Rainfall is characterised by its amount, intensity and distribution in time.

Amount of Rainfall

Imagine an open square container, 1 m wide, 1 m long and 0.5 m high.

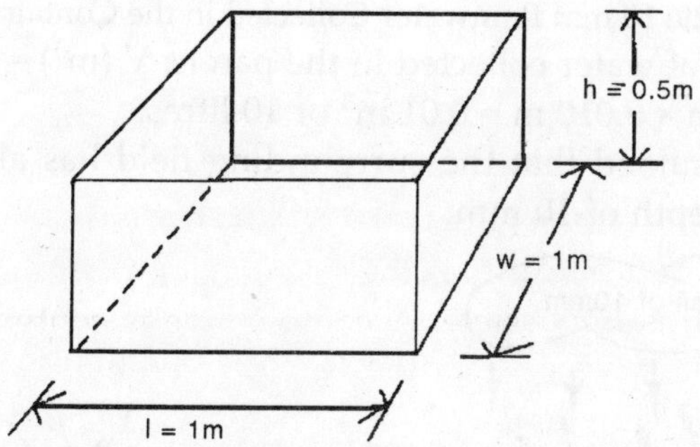

Fig:27. An Open Container to Collect Rainwater

This container is placed horizontally on an open area in a field.

Fig:28. Container Placed in the Field

During a rain shower, the container collects the water. Suppose that when the rain stops, the depth of water contained in the pan is 10 mm.

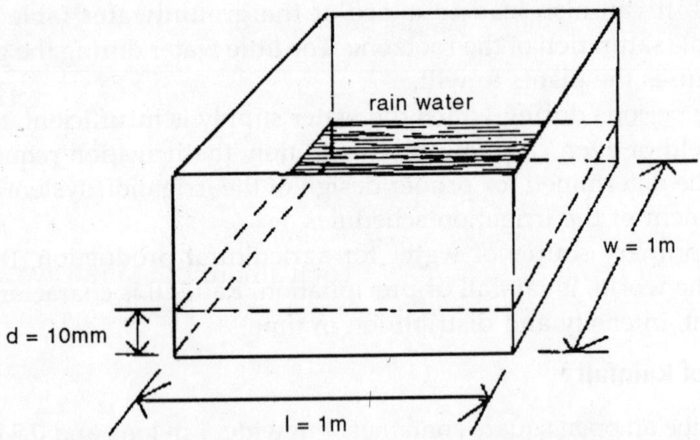

Fig:29. 10 mm Rainwater Collected in the Container

The volume of water collected in the pan is: V (m^3) = l (m) × w (m) × d (m) = 1 m × 1 m × 0.010 m = 0.01 m^3 or 10 litres.

It can be assumed that the surrounding field has also received an uniform water depth of 10 mm.

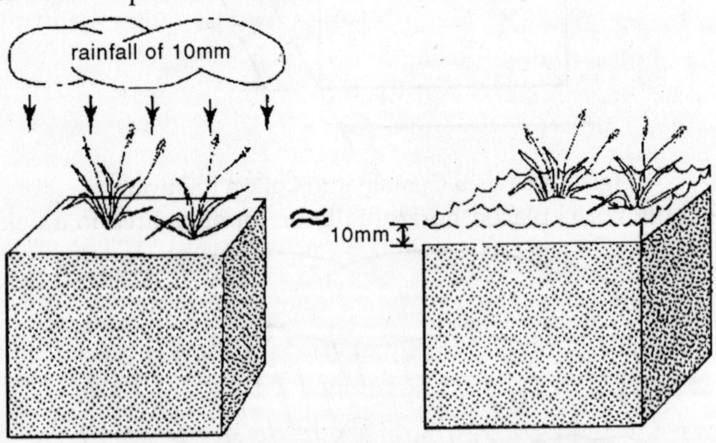

Fig:30. 10 mm Rainfall on the Field

In terms of volume, with a rainfall of 10 mm, every square metre of the field receives 0.01 m, or 10 litres, of rain water. With a rainfall of 1 mm, every square metre receives 1 litre of rain water. A rainfall of 1 mm supplies 0.001 m^3, or 1 litre of water to each square metre of the field. Thus 1 ha receives 10,000 litres.

Question: What is the total amount of water received by a field of 5 ha under a rainfall of 15 mm?

Answer: Each hectare (10,000 m^2) receives 10 000 m^2 × 0.015 m = 150 m^3 of water. Thus the total amount of water received by the 5 hectares is: 5 × 150 m. = 750 m.

Rainfall is often expressed in millimetres per day (mm/day) which represents the total depth of rainwater (mm), during 24 hours. It is the sum of all the rain showers which occurred during these 24 hours.

Rainfall Intensity

The rainfall intensity is the depth of water (in mm) received during a shower divided by the duration of the shower (in hours). It is expressed in millimetres of water depth per hour (mm/hour).

$$\text{Rainfallintensity (mm/hour)} = \frac{\text{Total amount of rain water (mm)}}{\text{Duration of the rainfall (hours)}}$$

For example, a rain shower lasts 3.5 hours and supplies 35 mm of water. The intensity of this shower is $\frac{35 \text{ mm}}{3.5 \text{ hours}} = 100$ mm/hour.

Suppose the same amount of water (35 mm) is supplied in one hour only, thus by a shower of higher intensity: $\frac{35 \text{ mm}}{1 \text{ hour}} = 35$ mm/hour.

Although the same amount of water (35 mm) has been supplied by both showers, the high intensity shower is less profitable to the crops. The high intensity rainfall usually has big drops that fall with more force on the soil surface.

In fine textured soil especially, the soil aggregates break down rapidly into fine particles that seal the soil surface. The infiltration is then reduced and surface runoff increases.

Fig:31. Sealing of the Soil Surface by Raindrops

The low intensity rainfall has finer drops. The soil surface is not sealed, the rainwater infiltrates more easily and surface runoff is limited.

Rainfall Distribution

Suppose that during one month, a certain area receives a total amount of rain water of 100 mm (100 mm/month). For crop growth, the distribution of the various showers during this month is important. Suppose that the rainwater falls during two showers of 50 mm each, one at the beginning of the month and the other one at the end of the month. In between these two showers, the crop undergoes a long dry period and may even wilt. Irrigation during this period is then required. On the other hand, if the rainwater is supplied regularly by little showers, evenly distributed over the month adequate soil moisture is continuously maintained and irrigation might not be required.

Not only the rainfall distribution within a month is important. It is also important to look into the rainfall distribution over the years. Suppose that in a certain area the average rainfall in May is 150 mm and that this amount is just sufficient to satisfy the water need of the crops during this month.

You may however find that, in this area, the rainfall in an exceptionally dry year is only 75 mm, while in a wet year the rainfall is 225 mm. In a dry year it would thus be necessary to irrigate the crops in May, while in an average year or a wet year, irrigation is not needed.

Effective Rainfall

When rain water falls on the soil surface, some of it infiltrates into the soil, some stagnates on the surface, while some flows over the surface as runoff. When the rainfall stops, some of the water stagnating on the surface evaporates to the atmosphere, while the rest slowly infiltrates into the soil. From all the water that infiltrates into the soil some percolates below the rootzone, while the rest remains stored in the rootzone.

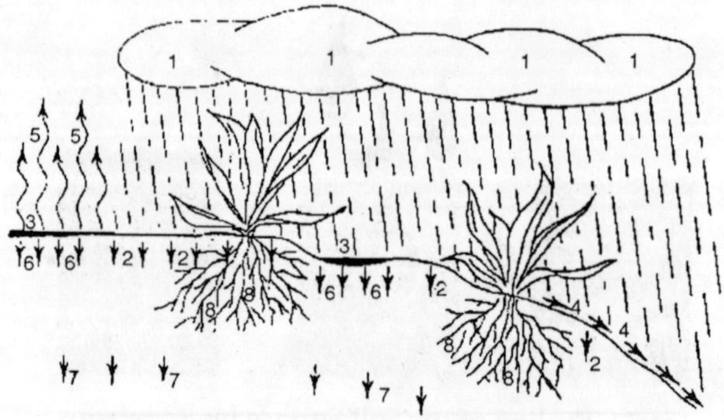

Fig:32. Effective Rainfall (8) = (1) – (4) – (5) – (7)

Factors Influencing Effective Rainfall

Many factors influence the amount of the effective rainfall. There are factors which the farmer cannot influence (e.g., the climate and the soil texture) and those which the farmer can influence (e.g., the soil structure).

Climate

The climate determines the amount, intensity and distribution of rainfall which have direct influence on the effective rainfall.

Soil Texture

In coarse textured soil, water infiltrates quickly but a large part of it percolates below the rootzone. In fine textured soil, the water infiltrates slowly, but much more water is kept in the rootzone than in coarse textured soil.

Soil Structure

The condition of the soil structure greatly influences the infiltration rate and therefore the effective rainfall. A favourable soil structure can be obtained by cultural practices (e.g., ploughing, mulching, ridging, etc.).

Depth of the Rootzone

Soil water stored in deep layers can be used by the plants only when roots penetrate to that depth. The depth of root penetration is primarily dependent on the type of crop, but also on the type of soil. The thicker the rootzone, the more water available to the plant.

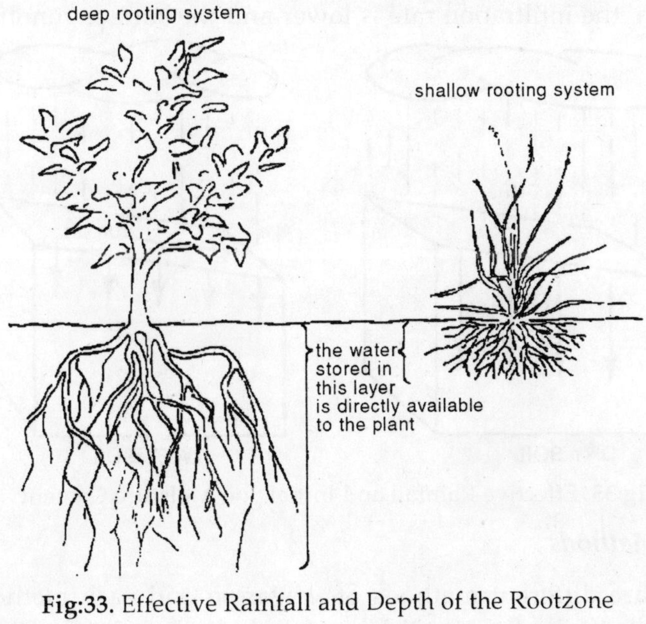

Fig:33. Effective Rainfall and Depth of the Rootzone

Topography

On steep sloping areas, because of high runoff, the water has less time to infiltrate than in rather flat areas. The effective rainfall is thus lower in sloping areas.

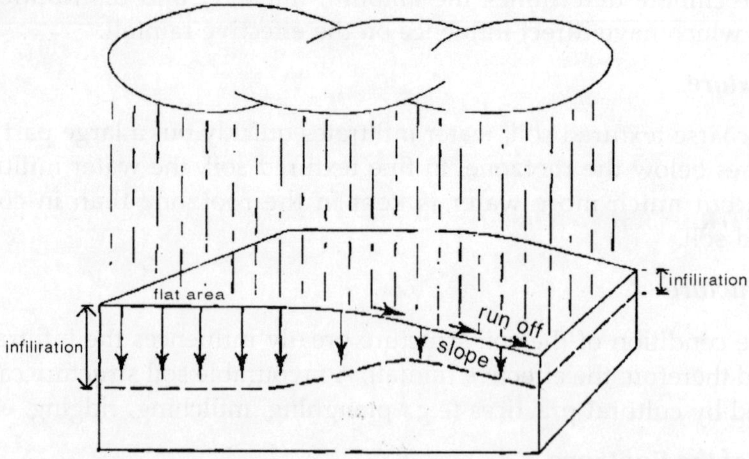

Fig:34. Effective Rainfall and Topography

Initial Soil Moisture Content

The infiltration rate is higher when soil is dry than when it is moist. This means that for a rain shower occurring shortly after a previous shower or irrigation, the infiltration rate is lower and the surface runoff higher.

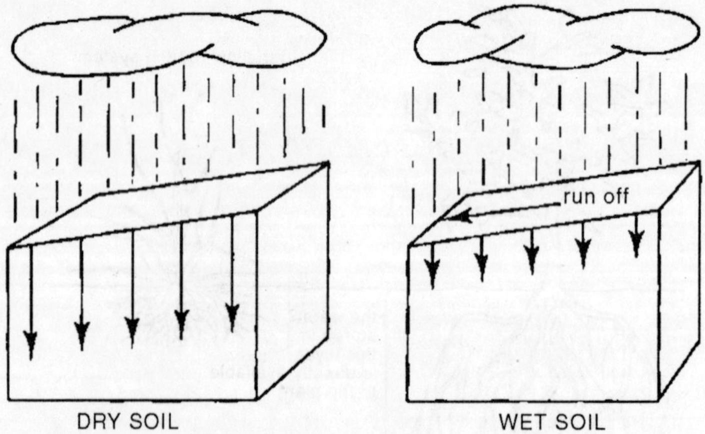

Fig:35. Effective Rainfall and Initial Soil Moisture Content

Irrigation Methods

There are different methods of irrigation and each method has a specific influence on the effective rainfall. In basin irrigation there is no surface runoff. All the rainwater is trapped in the basin and has time to infiltrate. In inclined border and furrow irrigation, the runoff is relatively large. At the lower end of the field the runoff water is collected in a field drain and carried away. Thus the effective rainfall under border or furrow irrigation is lower than under basin irrigation. In contour furrow irrigation there is very little or no slope in the direction of the furrow and thus runoff is limited; the runoff over the cross slope is also limited as the water is caught by the ridges. This results in a relatively high effective rainfall, compared to inclined border or furrow irrigation.

EVAPOTRANSPIRATION

Evaporation

Imagine the same open container as used for the collection of rain water, but this time with a depth of 10 mm of water in it; leave the container in the field for 24 hours. Make sure that it does not rain during those 24 hours.

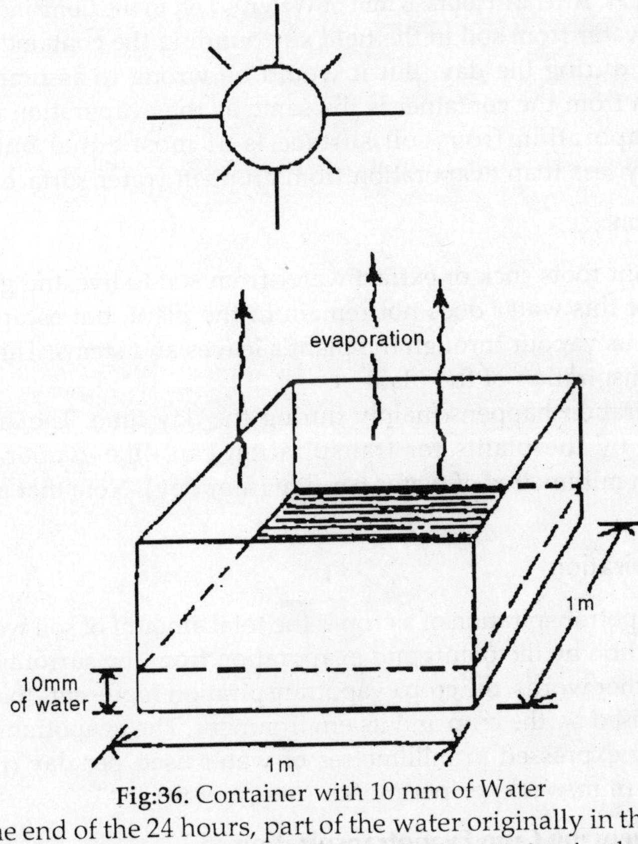

Fig:36. Container with 10 mm of Water

At the end of the 24 hours, part of the water originally in the container has evaporated. If only 6 mm of water depth remains in the container, then the evaporation during this day was 10 – 6 = 4 mm.

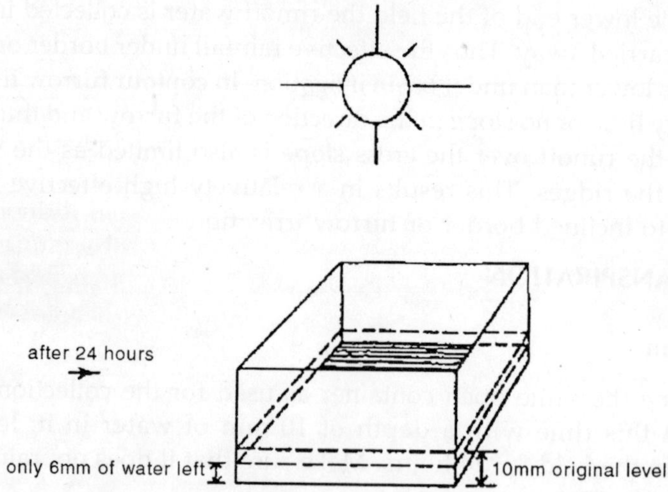

Fig:37. After 24 Hours, 6 mm of Water is Left in the Container

Some water from soil in the field surrounding the container has also evaporated during the day. But it would be wrong to assume that the evaporation from the container is the same as the evaporation from soil. In fact, evaporation from soil surface is at most equal but usually considerably less than evaporation from an open water surface.

Transpiration

The plant roots suck or extract water from soil to live and grow. The main part of this water does not remain in the plant, but escapes to the atmosphere as vapour through the plant's leaves and stems. This process is called transpiration of the plant.

Transpiration happens mainly during the day time. The amount of water used by the plants for transpiration can, like evaporation, be expressed in millimetres of water per day (mm/day). Note that a day has 24 hours.

Evapotranspiration

The evapotranspiration of a crop is the total amount of soil water used for transpiration by the plants and evaporation from the surrounding soil surface. In other words, the crop evapotranspiration represents the amount of water utilised by the crop and its environment. The evapotranspiration is commonly expressed in millimetres of water used per day (mm/day) or per week (mm/week) or per month (mm/month).

Factors Influencing Crop Evapotranspiration

Many factors influence the evapotranspiration of the crop.

Factor	*Effect on Crop Evapotranspiration*	
	High	*Low*
Climate	hot	cool
	dry	wet
	windy	no wind
	no clouds	cloudy
Crop	mid/late season	initial or ripening
	dense plant spacing	wide plant spacing
Soil moisture	moist	dry

IRRIGATION SYSTEM

The irrigation system consists of a (main) intake structure or (main) pumping station, a conveyance system, a distribution system, a field application system and a drainage system.

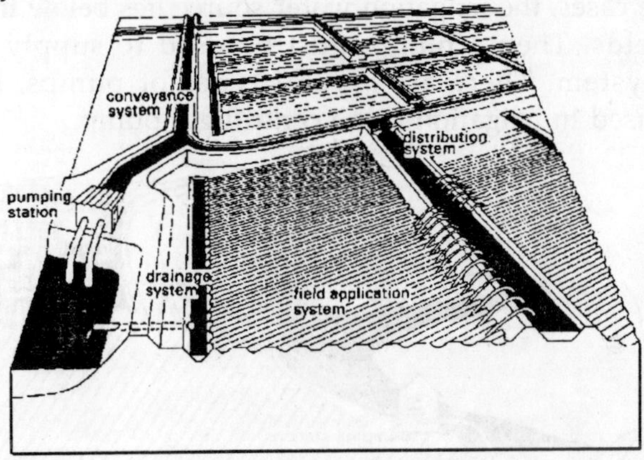

Fig:38. An Irrigation System

The (main) intake structure, or (main) pumping station, directs water from the source of supply, such as a reservoir or a river, into the irrigation system.

The conveyance system assures the transport of water from the main intake structure or main pumping station up to the field ditches. The distribution system assures the transport of water through field ditches to the irrigated fields. The field application system assures the transport of water within the fields. The drainage system removes the excess water (caused by rainfall and/or irrigation) from the fields.

Main Intake Structure and Pumping Station

Main Intake Structure

The intake structure is built at the entry to the irrigation system. Its purpose is to direct water from the original source of supply (lake, river, reservoir, etc.) into the irrigation system.

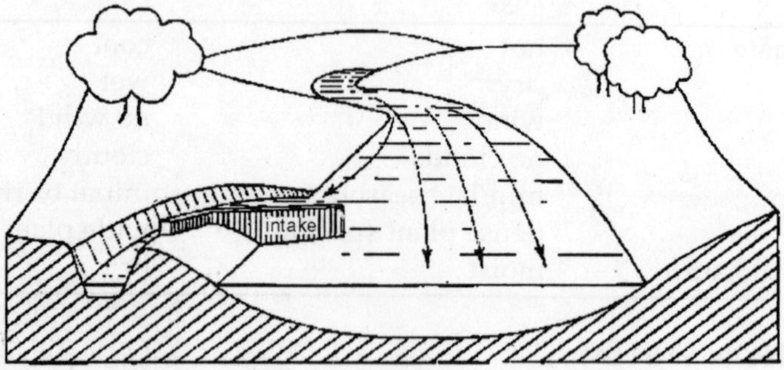

Fig:39. An Intake Structure

Pumping Station

In some cases, the irrigation water source lies below the level of the irrigated fields. Then a pump must be used to supply water to the irrigation system. There are several types of pumps, but the most commonly used in irrigation is the centrifugal pump.

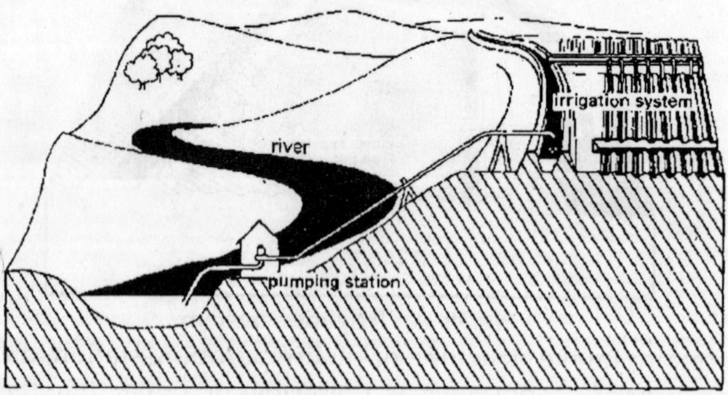

Fig:40. A Pumping Station

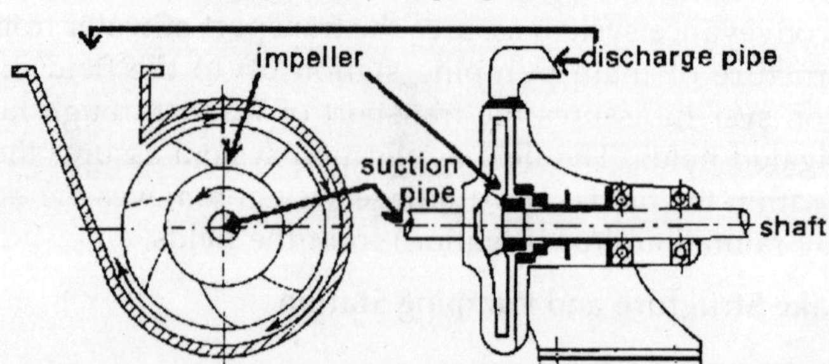

Fig:41. Diagram of a Centrifugal Pump

The centrifugal pump consists of a case in which an element, called an impeller, rotates driven by a motor. Water enters the case at the centre, through the suction pipe. The water is immediately caught by the rapidly rotating impeller and expelled through the discharge pipe. The centrifugal pump will only operate when the case is completely filled with water.

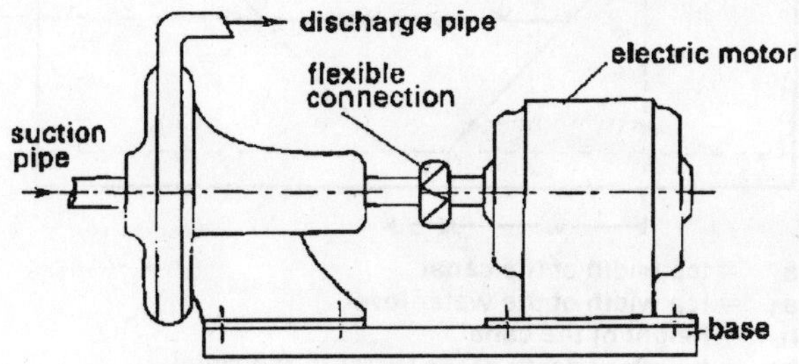

Fig:42. Centrifugal Pump and Motor

Conveyance and Distribution System

The conveyance and distribution systems consist of canals transporting the water through the whole irrigation system. Canal structures are required for the control and measurement of the water flow.

Open Canals

An open canal, channel, or ditch, is an open waterway whose purpose is to carry water from one place to another. Channels and canals refer to main waterways supplying water to one or more farms. Field ditches have smaller dimensions and convey water from the farm entrance to the irrigated fields.

Canal Characteristics

According to the shape of their cross-section, canals are called rectangular (a), triangular (b), trapezoidal (c), circular (d), parabolic (e) and irregular or natural (f).

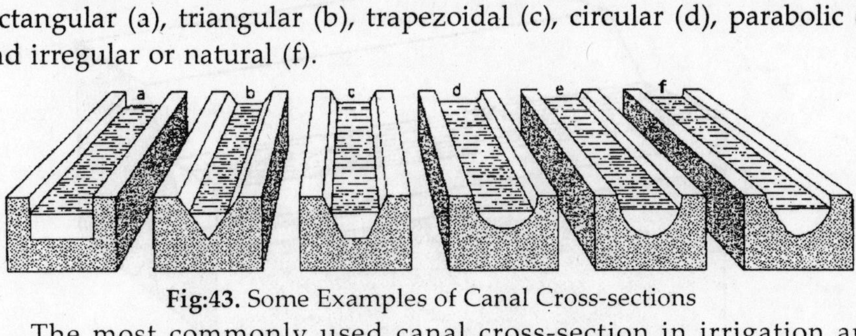

Fig:43. Some Examples of Canal Cross-sections

The most commonly used canal cross-section in irrigation and drainage, is the trapezoidal cross-section. For the purposes of this publication, only this type of canal will be considered. The typical cross-section of a trapezoidal canal.

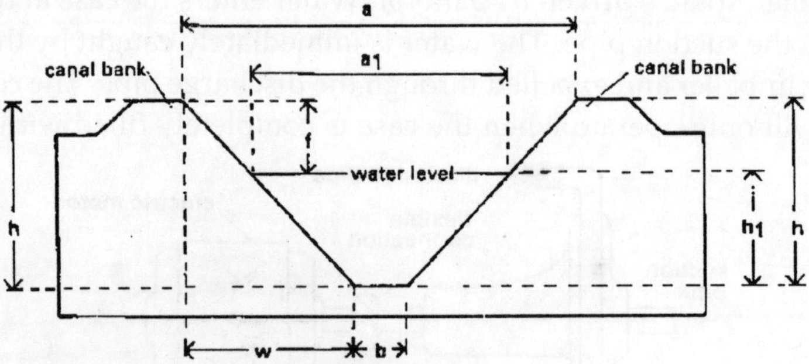

a = top width of the canal
a_1 = top width of the water level
h = height of the canal
h_1 = heigth or depth of the water in the canal
b = bottom width of the canal
h:w = side slope of the canal
f = free board (= $h-h_1$)

Fig:44. A Trapezoidal Canal Cross-section

The freeboard of the canal is the height of the bank above the highest water level anticipated. It is required to guard against overtopping by waves or unexpected rises in the water level.

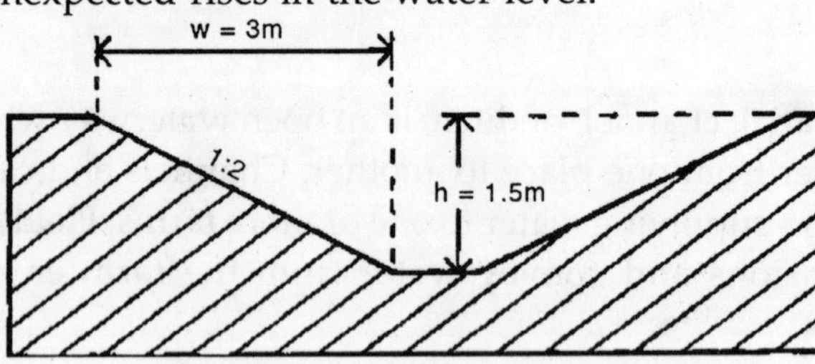

side slope: h:w = 1.5:3 = 1:2 (one to two)

Fig:45. A Side Slope of 1:2 (One to Two)

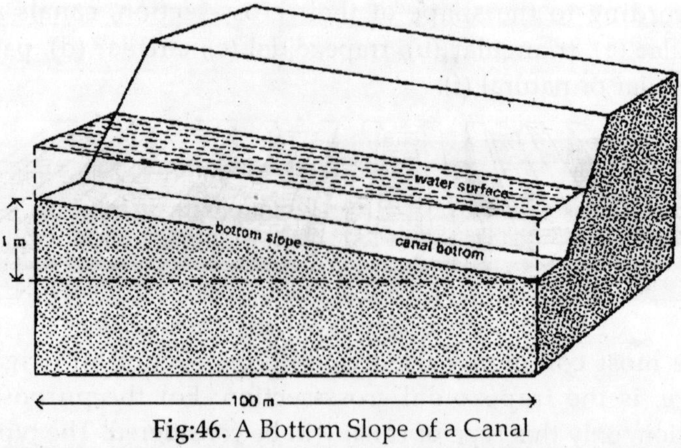

Fig:46. A Bottom Slope of a Canal

The bottom slope of the canal does not appear on the drawing of the cross-section but on the longitudinal section. It is commonly expressed in per cent or per mil. The side slope of the canal is expressed as ratio, namely, the vertical distance or height to the horizontal distance or width. For example, if the side slope of the canal has a ratio of 1:2 (one to two), this means that the horizontal distance (w) is two times the vertical distance (h).

An example of the calculation of the bottom slope of a canal is given below:

$$\text{The bottom slope (per cent)} = \frac{\text{height difference (metres)}}{\text{horizontal distance (metres)}} \times 100$$

$$= \frac{1\,\text{m}}{100\,\text{m}} \times 100 = 1\%$$

or

$$\text{The bottom slope (per cent)} = \frac{\text{height difference (metres)}}{\text{horizontal distance (metres)}} \times 1000$$

$$= \frac{1\,\text{m}}{100\,\text{m}} \times 1000 = 10‰$$

Earthen Canals

Earthen canals are simply dug in the ground and the bank is made up from the removed earth. The disadvantages of earthen canals are the risk of the side slopes collapsing and the water loss due to seepage. They also require continuous maintenance in order to control weed growth and to repair damage done by livestock and rodent.

Lined Canals

Earthen canals can be lined with impermeable materials to prevent excessive seepage and growth of weeds. Lining canals is also an effective way to control canal bottom and bank erosion. The materials mostly used for canal lining are concrete (in precast slabs or cast in place), brick or rock masonry and asphaltic concrete (a mixture of sand, gravel and asphalt). The construction cost is much higher than for earthen canals. Maintenance is reduced for lined canals, but skilled labour is required.

Canal Structures

The flow of irrigation water in the canals must always be under control. For this purpose, canal structures are required. They help regulate the flow and deliver the correct amount of water to the different branches of the system and onward to the irrigated fields. There are four main types of structures: erosion control structures, distribution control structures, crossing structures and water measurement structures.

Erosion Control Structures

Canal Erosion

Canal bottom slope and water velocity are closely related, as the following example will show.

A cardboard sheet is lifted on one side 2 cm from the ground. A small ball is placed at the edge of the lifted side of the sheet. It starts rolling downward, following the slope direction. The sheet edge is now lifted 5 cm from the ground creating a steeper slope. The same ball placed on the top edge of the sheet rolls downward, but this time much faster. The steeper the slope, the higher the velocity of the ball.

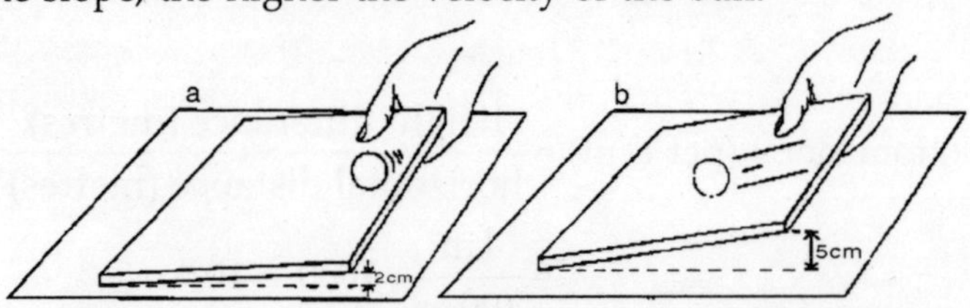

Fig:47. The Relationship between Slope and Velocity

Water poured on the top edge of the sheet reacts exactly the same as the ball. It flows downward and the steeper the slope, the higher the velocity of the flow. Water flowing in steep canals can reach very high velocities. Soil particles along the bottom and banks of an earthen canal are then lifted, carried away by the water flow and deposited downstream where they may block the canal and silt up structures. The canal is said to be under erosion; the banks might eventually collapse.

Drop Structures and Chutes

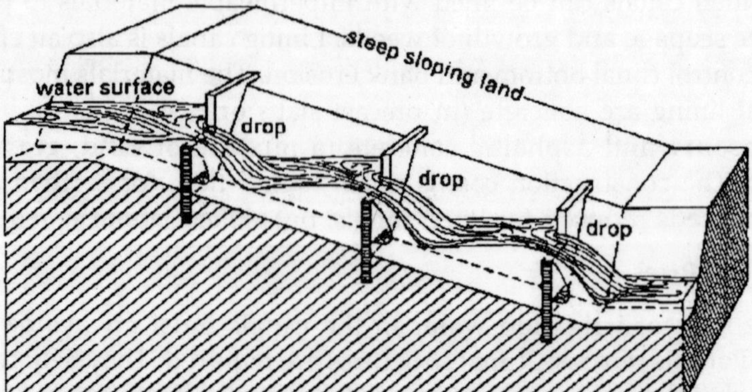

Fig:48. Longitudinal Section of a Series of Drop Structures

Drop structures or chutes are required to reduce the bottom slope of canals lying on steeply sloping land in order to avoid high velocity of the flow and risk of erosion. These structures permit the canal to be constructed as a series of relatively flat sections.

Drop structures take the water abruptly from a higher section of the canal to a lower one. In a chute, the water does not drop freely but is carried through a steep, lined canal section. Chutes are used where there are big differences in the elevation of the canal.

Distribution Control Structures

Distribution control structures are required for easy and accurate water distribution within the irrigation system and on the farm. Division boxes are used to divide or direct the flow of water between two or more canals or ditches.

Water enters the box through an opening on one side and flows out through openings on the other sides. These openings are equipped with gates.

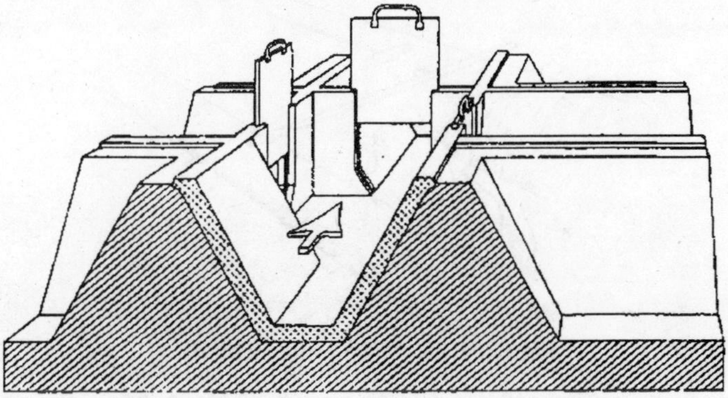

Fig:49. A Division Box with Three Gates

Turnouts

Turnouts are constructed in the bank of a canal. They divert part of the water from the canal to a smaller one.

Turnouts can be concrete structures, or pipe structures.

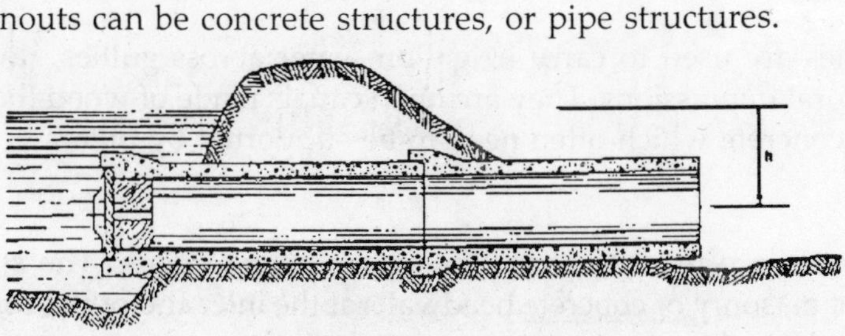

Fig:50. A Pipe Turnout

Checks

To divert water from the field ditch to the field, it is often necessary to raise the water level in the ditch. Checks are structures placed across the ditch to block it temporarily and to raise the upstream water level. Checks can be permanent structures or portable.

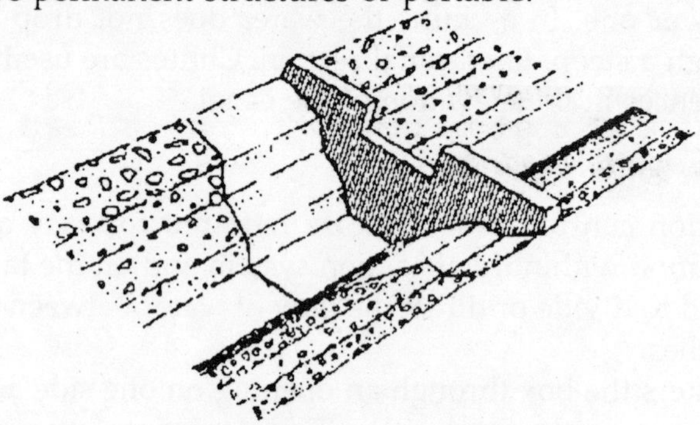

Fig:51. A Permanent Concrete Check

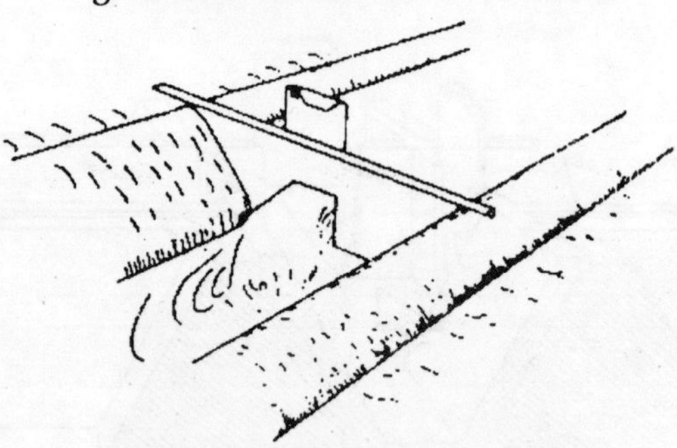

Fig:52. A Portable Metal Check

Crossing Structures

It is often necessary to carry irrigation water across roads, hillsides and natural depressions. Crossing structures, such as flumes, culverts and inverted siphons, are then required.

Flumes

Flumes are used to carry irrigation water across gullies, ravines or other natural depressions. They are open canals made of wood (bamboo), metal or concrete which often need to be supported by pillars.

Culverts

Culverts are used to carry the water across roads. The structure consists of masonry or concrete headwalls at the inlet and outlet connected by a buried pipeline.

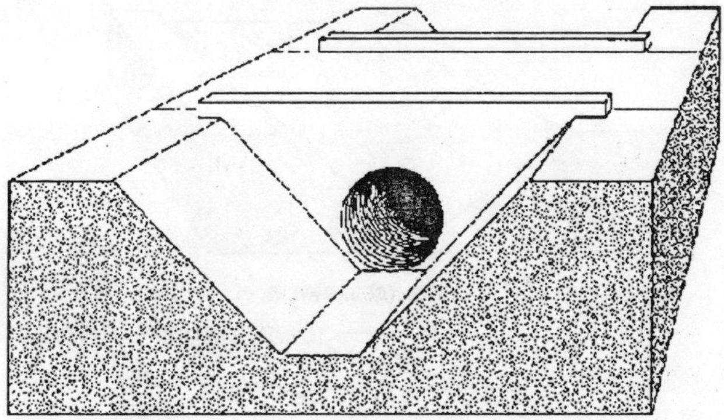

Fig:53. A Culvert

Inverted Siphons

When water has to be carried across a road which is at the same level as or below the canal bottom, an inverted siphon is used instead of a culvert. The structure consists of an inlet and outlet connected by a pipeline. Inverted siphons are also used to carry water across wide depressions.

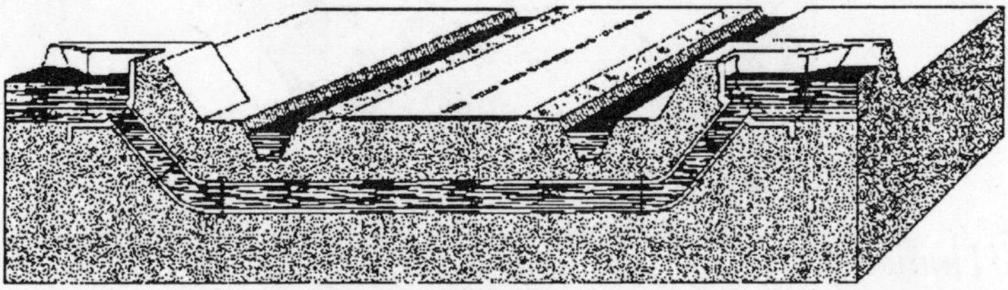

Fig:54. An Inverted Siphon

Water Measurement Structures

The principal objective of measuring irrigation water is to permit efficient distribution and application. By measuring the flow of water, a farmer knows how much water is applied during each irrigation. In irrigation schemes where water costs are charged to the farmer, water measurement provides a basis for estimating water charges. The most commonly used water measuring structures are weirs and flumes. In these structures, the water depth is read on a scale which is part of the structure. Using this reading, the flow-rate is then computed from standard formulas or obtained from standard tables prepared specially for the structure.

Weirs

In its simplest form, a weir consists of a wall of timber, metal or concrete with an opening with fixed dimensions cut in its edge. The opening, called a notch, may be rectangular, trapezoidal or triangular.

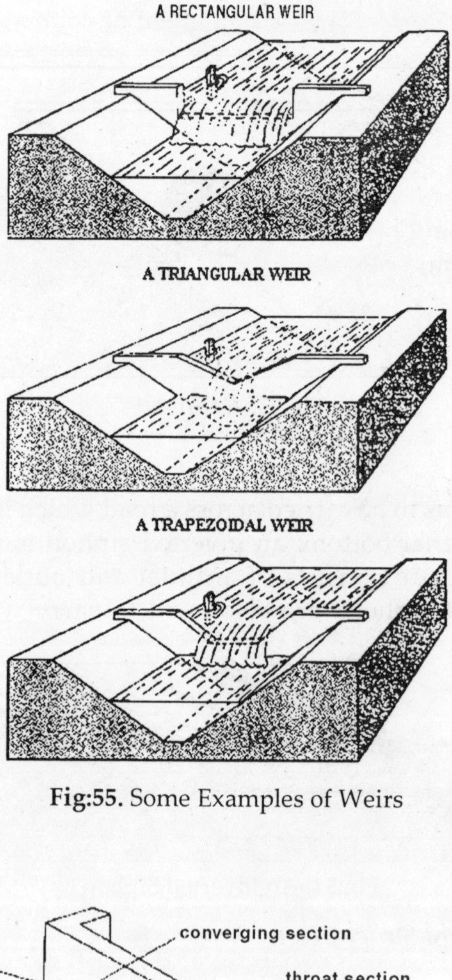

Fig:55. Some Examples of Weirs

Parshall Flumes

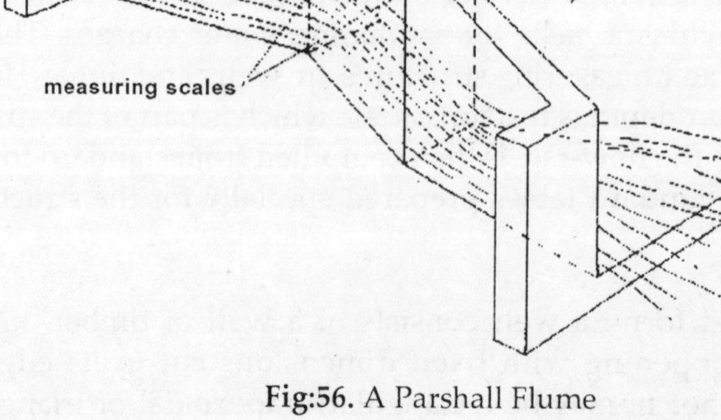

Fig:56. A Parshall Flume

The Parshall flume consists of a metal or concrete channel structure with three main sections:

- A converging section at the upstream end, leading to;
- A constricted or throat section,
- A diverging section at the downstream end.

Depending on the flow condition (free flow or submerged flow), the water depth readings are taken on one scale only (the upstream one) or on both scales simultaneously.

Cut-throat Flume

The cut-throat flume is similar to the Parshall flume, but has no throat section, only converging and diverging sections. Unlike the Parshall flume, the cut-throat flume has a flat bottom. Because it is easier to construct and install, the cut-throat flume is often preferred to the Parshall flume.

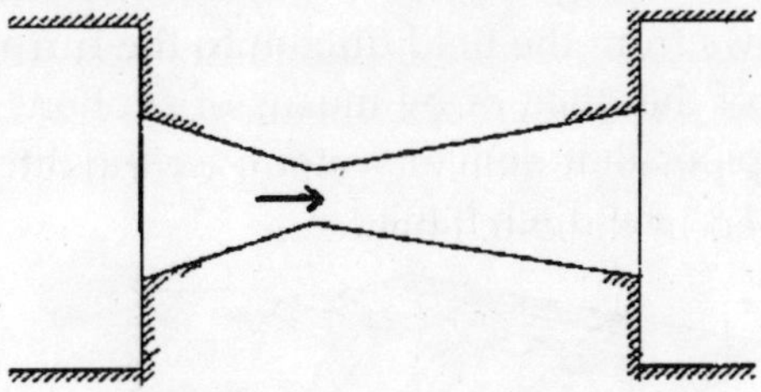

Fig:57. A Cut-throat Flume

Field Application Systems

There are many methods of applying water to the field. The simplest one consists of bringing water from the source of supply, such as a well, to each plant with a bucket or a water-can.

Fig:58. Watering Plants with a Bucket

This is a very time-consuming method and it involves quite heavy work. However, it can be used successfully to irrigate small plots of land, such as vegetable gardens, that are in the neighbourhood of a water source. More sophisticated methods of water application are used in larger irrigation systems. There are three basic methods: surface irrigation, sprinkler irrigation and drip irrigation.

Surface Irrigation

Surface irrigation is the application of water to the fields at ground level. Either the entire field is flooded or the water is directed into furrows or borders.

Furrow Irrigation

Furrows are narrow ditches dug on the field between the rows of crops. The water runs along them as it moves down the slope of the field.

The water flows from the field ditch into the furrows by opening up the bank or dyke of the ditch or by means of syphons or spiles. Siphons are small curved pipes that deliver water over the ditch bank. Spiles are small pipes buried in the ditch bank.

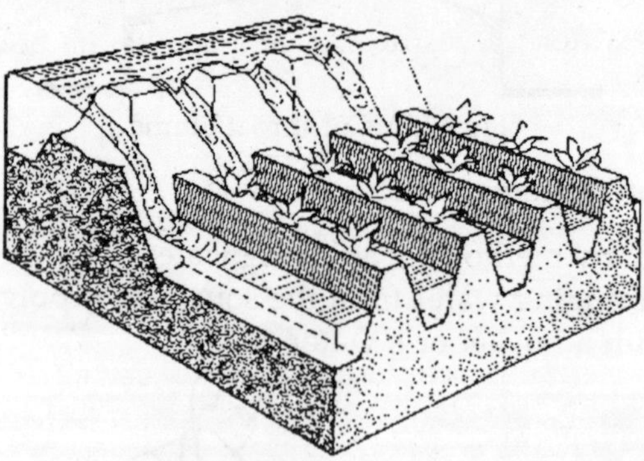

Fig:59. Water Flows into the Furrows Through Openings in the Bank

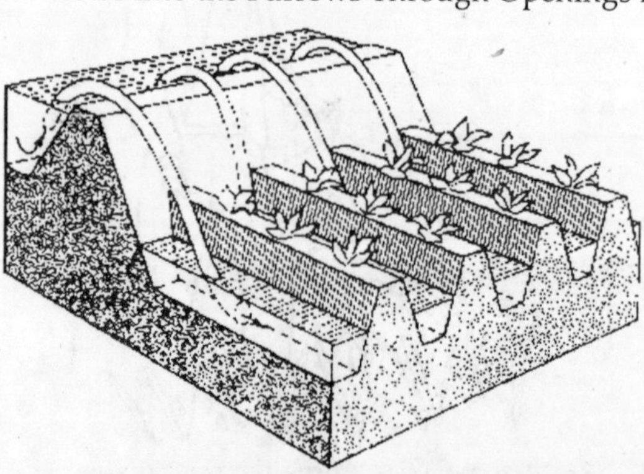

Fig:60. The Use of Siphons

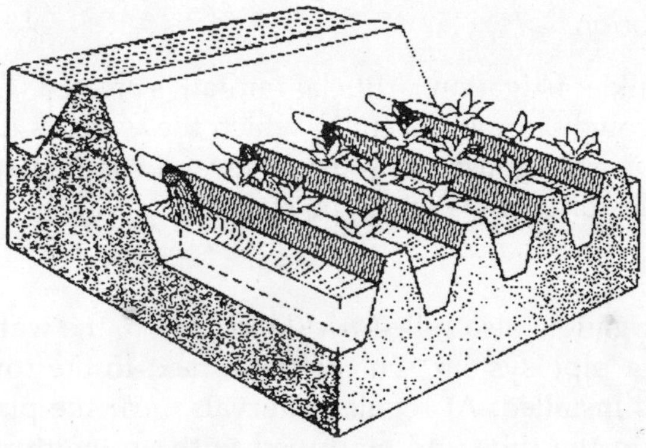

Fig:61. The Use of Spiles

Border Irrigation

In border irrigation, the field to be irrigated is divided into strips (also called borders or borderstrips) by parallel dykes or border ridges. The water is released from the field ditch onto the border through gate structures called outlets.

The water can also be released by means of siphons or spiles. The sheet of flowing water moves down the slope of the border, guided by the border ridges.

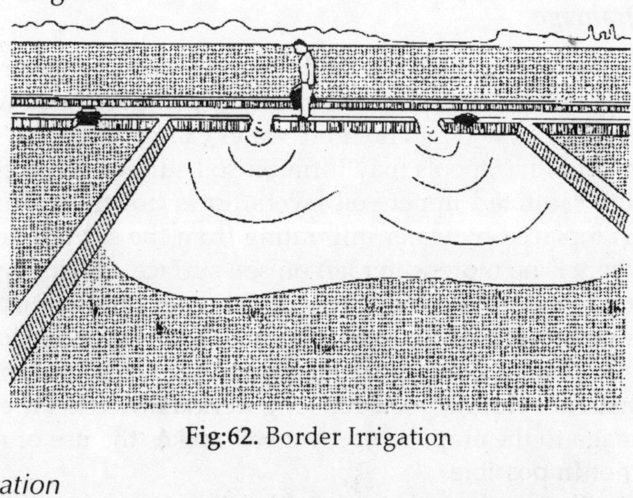

Fig:62. Border Irrigation

Basin Irrigation

Basins are horizontal, flat plots of land, surrounded by small dykes or bunds. The banks prevent the water from flowing to the surrounding fields. Basin irrigation is commonly used for rice grown on flat lands or in terraces on hillsides.

Trees can also be grown in basins, where one tree usually is located in the centre of a small basin.

Sprinkler Irrigation

With sprinkler irrigation, artificial rainfall is created. The water is led to the field through a pipe system in which the water is under pressure. The spraying is accomplished by using several rotating sprinkler heads or spray nozzles or a single gun type sprinkler.

Drip Irrigation

In drip irrigation, also called trickle irrigation, the water is led to the field through a pipe system. On the field, next to the row of plants or trees, a tube is installed. At regular intervals, near the plants or trees, a hole is made in the tube and equipped with an emitter. The water is supplied slowly, drop by drop, to the plants through these emitters.

Drainage System

A drainage system is necessary to remove excess water from the irrigated land. This excess water may be e.g., waste water from irrigation or surface runoff from rainfall. It may also include leakage or seepage water from the distribution system.

Drainage

Need for Drainage

During rain or irrigation, the fields become wet. The water infiltrates into soil and is stored in its pores. When all the pores are filled with water, soil is said to be saturated and no more water can be absorbed; when rain or irrigation continues, pools may form on soil surface. Part of the water present in the saturated upper soil layers flows downward into deeper layers and is replaced by water infiltrating from the surface pools.

When there is no more water left on soil surface, the downward flow continues for a while and air re-enters in the pores of soil. This soil is not saturated anymore. However, saturation may have lasted too long for the plants' health. Plant roots require air as well as water and most plants cannot withstand saturated soil for long periods (rice is an exception). Besides damage to the crop, a very wet soil makes the use of machinery difficult, if not impossible.

The water flowing from the saturated soil downward to deeper layers, feeds the groundwater reservoir. As a result, the groundwater level rises. Following heavy rainfall or continuous over-irrigation, the groundwater table may even reach and saturate part of the rootzone.

Again, if this situation lasts too long, the plants may suffer. Measures to control the rise of the water table are thus necessary. The removal of excess water either from the ground surface or from the rootzone, is called drainage. Excess water may be caused by rainfall or by using too much irrigation water, but may also have other origins such as canal seepage or floods.

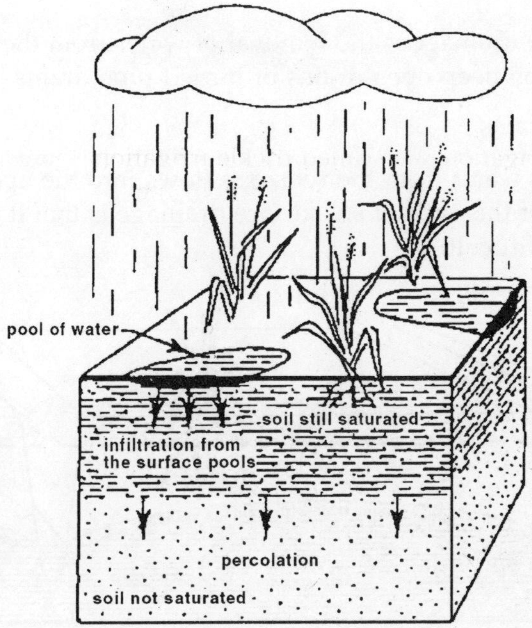

Fig:63. Water Percolates to Deeper Layers and Infiltrates from the Pools

In very dry areas there is often accumulation of salts in soil. Most crops do not grow well on salty soil. Salts can be washed out by percolating irrigation water through the rootzone of the crops. To achieve sufficient percolation, farmers will apply more water to the field than the crops need. But the salty percolation water will cause the water table to rise. Drainage to control the water table, therefore, also serves to control the salinity of soil.

Different Types of Drainage

Drainage can be either natural or artificial. Many areas have some natural drainage; this means that excess water flows from the farmers' fields to swamps or to lakes and rivers. Natural drainage, however, is often inadequate and artificial or man-made drainage is required.

There are two types of artificial drainage: surface drainage and subsurface drainage.

Surface Drainage

Surface drainage is the removal of excess water from the surface of the land. This is normally accomplished by shallow ditches, also called open drains. The shallow ditches discharge into larger and deeper collector drains. In order to facilitate the flow of excess water toward the drains, the field is given an artificial slope by means of land grading.

Subsurface Drainage

Subsurface drainage is the removal of water from the rootzone. It is accomplished by deep open drains or buried pipe drains.

Deep Open Drains

The excess water from the rootzone flows into the open drains. The disadvantage of this type of subsurface drainage is that it makes the use of machinery difficult.

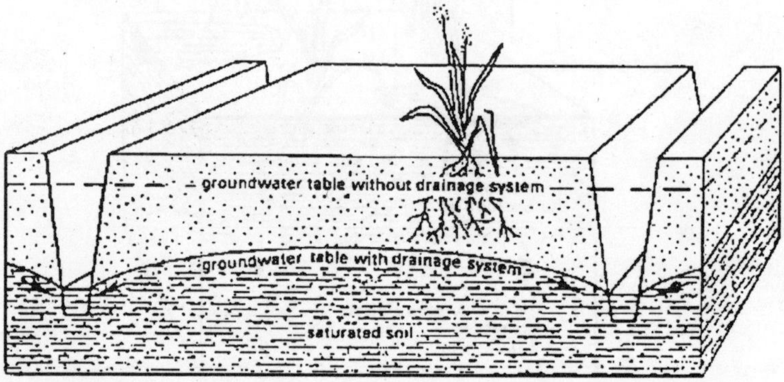

Fig:64. Control of the Groundwater of Deep Open Drains

Pipe Drains

Pipe drains are buried pipes with openings through which the soil water can enter. The pipes convey the water to a collector drain.

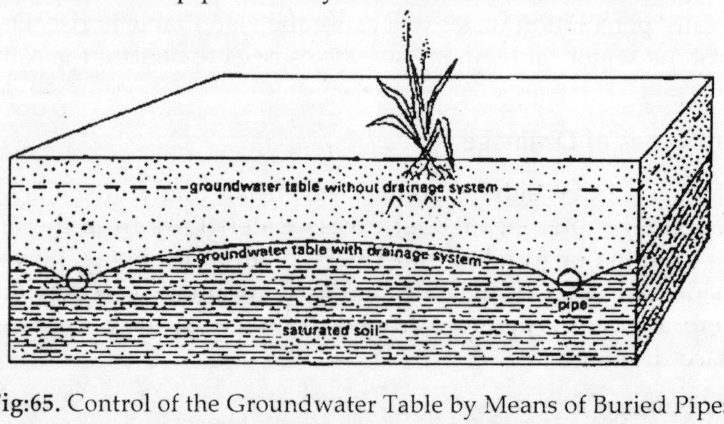

Fig:65. Control of the Groundwater Table by Means of Buried Pipes

Drain pipes are made of clay, concrete or plastic. They are usually placed in trenches by machines. In clay and concrete pipes (usually 30 cm long and 5 - 10 cm in diameter) drainage water enters the pipes through the joints. Flexible plastic drains are much longer (up to 200 m) and the water enters through perforations distributed over the entire length of the pipe.

Open drains use land that otherwise could be used for crops. They restrict the use of machines. They also require a large number of bridges and culverts for road crossings and access to the fields. Open drains require frequent maintenance (weed control, repairs, etc.).

In contrast to open drains, buried pipes cause no loss of cultivable land and maintenance requirements are very limited. The installation costs, however, of pipe drains may be higher due to the materials, the equipment and the skilled manpower involved.

SOIL SALINISATION

A soil may be rich in salts because the parent rock from which it was formed contains salts. Sea water is another source of salts in low-lying areas along the coast. A very common source of salts in irrigated soils is the irrigation water itself.

Most irrigation waters contain some salts. After irrigation, the water added to the soil is used by the crop or evaporates directly from the moist soil. The salt, however, is left behind in the soil. If not removed, it accumulates in the soil; this process is called salinisation. Very salty soils are sometimes recognisable by a white layer of dry salt on the soil surface.

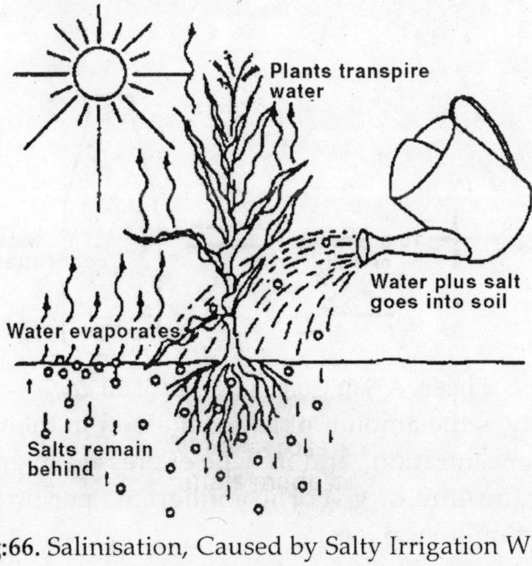

Fig:66. Salinisation, Caused by Salty Irrigation Water

Salty groundwater may also contribute to salinisation. When the water table rises (e.g., following irrigation in the absence of proper drainage), the salty groundwater may reach the upper soil layers and, thus, supply salts to the rootzone.

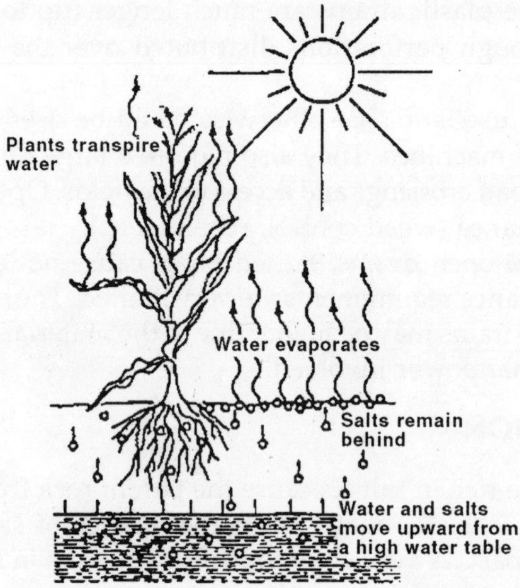

Fig:67. Salinisation, Caused by a High

Soils that contain a harmful amount of salt are often referred to as salty or saline soils. Soil, or water, that has a high content of salt is said to have a high salinity.

Salinity

Water Salinity

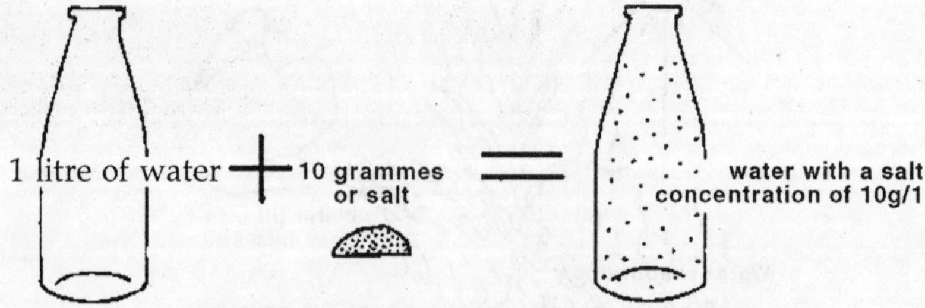

Fig:68. A Salt Concentration of 10 g/l

Water salinity is the amount of salt contained in the water. It is also called the "salt concentration" and may be expressed in grams of salt per litre of water (grams/litre or g/l) or in milligrams per litre (which is the same as parts per million, p.p.m).

However, the salinity of both water and soil is easily measured by means of an electrical device. It is then expressed in terms of electrical conductivity: millimhos/cm or micromhos/cm.

A salt concentration of 1 grammes per litre is about 1.5 millimhos/

cm. Thus a concentration of 3 grammes per litre will be about the same as 4.5 millimhos/cm.

Soil Salinity

The salt concentration in the water extracted from a saturated soil (called saturation extract) defines the salinity of this soil. If this water contains less than 3 grammes of salt per litre, the soil is said to be non saline. If the salt concentration of the saturation extract contains more than 12 g/l, the soil is said to be highly saline.

Salt Concentration of the Soil Water (saturation extract)		Salinity
in g/l	in millimhos/cm	
0 - 3	0 - 4.5	non saline
3 - 6	4.5 - 9	slightly saline
6 - 12	9 - 18	medium saline
more than 12	more than 18	highly saline

Crops and Saline Soils

Most crops do not grow well on soils that contain salts. One reason is that salt causes a reduction in the rate and amount of water that the plant roots can take up from the soil. Also, some salts are toxic to plants when present in high concentration.

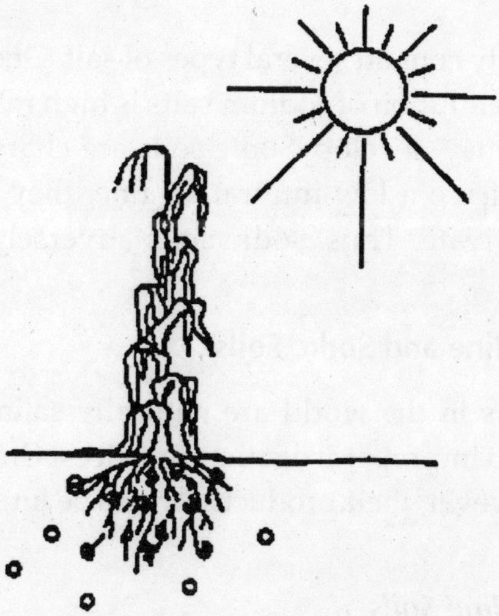

Fig:69. A High Salt Concentration in the Soil is Harmful for the Plants as the Water Uptake is Reduced

The highly tolerant crops can withstand a salt concentration of the saturation extract up to 10 g/l. The moderately tolerant crops can withstand salt concentration up to 5 g/l. The limit of the sensitive group is about 2.5 g/l.

Some plants are more tolerant to a high salt concentration than others:

Highly Tolerant	*Moderately Tolerant*	*Sensitive*
Date palm	Wheat	Red clover
Barley	Tomato	Peas
Sugarbeet	Oats	Beans
Cotton	Alfalfa	Sugarcane
Asparagus	Rice	Pear
Spinach	Maize	Apple
	Flax	Orange
	Potatoes	Prune
	Carrot	Plum
	Onion	Almond
	Cucumber	Apricot
	Pomegranate	Peach
	Fig	
	Olive	
	Grape	

Sodicity

Salty soils usually contain several types of salt. One of these is sodium salt. Where the concentration of sodium salts is high relative to other types of salt, a sodic soil may develop. Sodic soils are characterised by a poor soil structure: they have a low infiltration rate, they are poorly aerated and difficult to cultivate. Thus, sodic soils adversely affect the plants' growth.

Improvement of Saline and Sodic Soils

Numerous areas in the world are naturally saline or sodic or have become saline due to improper irrigation practices. Crop growth on many of these is poor. However, their productivity can be improved by a number of measures.

Improvement of Saline Soils

Improvement of a saline soil implies the reduction of the salt concentration of soil to a level that is not harmful to the crops.

To that end, more water is applied to the field than is required for crop growth. This additional water infiltrates into the soil and percolates through the rootzone. During percolation, it takes up part of the salts in the soil and takes these along to deeper soil layers. In fact, the water washes the salts out of the rootzone. This washing process is called leaching.

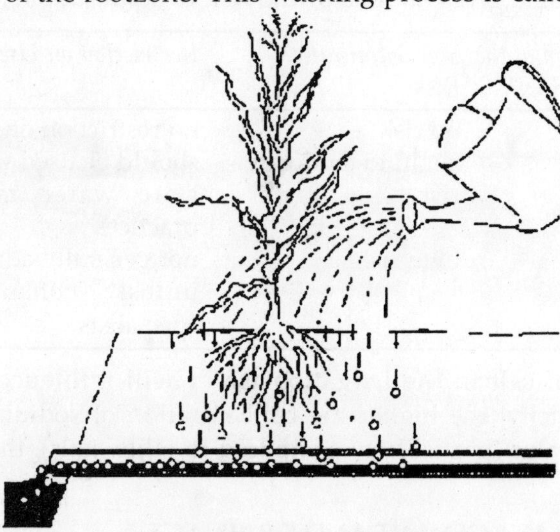

Fig:70. Leaching of Salts

The additional water required for leaching must be removed from the rootzone by means of a subsurface drainage system. If not removed, it could cause a rise of the groundwater table which would bring the salts back into the rootzone. Thus, improvement of saline soils includes, essentially, leaching and sub-surface drainage.

Improvement of Sodic Soils

Improvement of sodic soils implies the reduction of the amount of sodium present in the soil. This is done in two stages. Firstly, chemicals (such as gypsum), which are rich in calcium, are mixed with the soil; the calcium replaces the sodium. Then, the replaced sodium is leached from the rootzone by irrigation water.

Prevention of Salinization

Soils will become salty if salts are allowed to accumulate. Proper irrigation management and adequate drainage are not only important measures for the improvement of salty soils, they are also essential for the prevention of salinization.

Irrigation Water Quality

The suitability of water for irrigation depends on the amount and the type of salt the irrigation water contains. The higher the salt concentration of the irrigation water, the greater the risk of salinisation. The following table gives an idea of the risk of salinisation:

Salt Concentration of the Irrigation Water in g/l	*Soil Salinisation Risk*	*Restriction on Use*
less than 0.5 g/l	no risk	no restriction on its use
0.5 - 2 g/l	slight to moderate risk	should be used with appropriate water management practices
more than 2 g/l	high risk	not generally advised for use unless consulted with specialists

The type of salt in the irrigation water will influence the risk of developing sodicity: the higher the concentration of sodium present in the irrigation water (particularly compared to other soils), the higher the risk.

IRRIGATION MANAGEMENT AND DRAINAGE

Irrigation systems are never fully efficient. Some water is always lost in canals and on the farmers' fields. Part of this seeps into soil. While this will help leach salt out of the rootzone, it will also contribute to a rise of the water table; a high water table is risky because it may cause the salts to return to the rootzone.

Therefore, both the water losses and the water table must be strictly controlled. This requires careful management of the irrigation system and a good subsurface drainage system.

Chapter 6

Soil Vegetation

Water is probably our least understood natural resource. The earth has virtually the same amount of water today as it did when dinosaurs roamed the planet. Water covers nearly three-fourths of the earth as rivers, lakes and oceans, but only about 3 per cent of the planet's water is fresh, and two-thirds of that is ice. About 0.6 per cent of the earth's water is in the earth's underground aquifers, and a small but very important amount (0.003 per cent) is contained in plants, animals and soil. Over four trillion gallons of water fall on the United States daily in the form of precipitation, but much of that disappears in evaporation or runoff. The amount that soaks into soil determines, in part, plant life and productivity.

The hydrologic cycle is the continuous process by which water is transported from the oceans to the atmosphere to the land and back to the sea. Water evaporates from water bodies such as oceans, ponds and rivers and is moved across the earth as water vapour by wind currents. Soil, plants, people, animals, factories and vehicles also contribute to this vapour. Water vapour condenses and falls to earth as rain, sleet, snow or hail depending on the region, topography, climate and season.

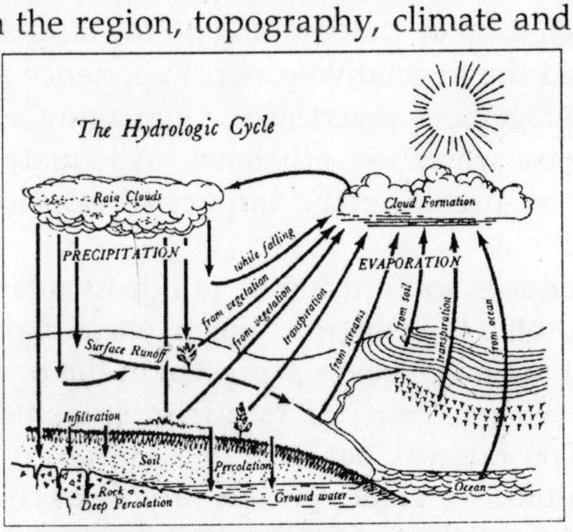

Fig:1. The Hydrologic Cycle

A large portion of the precipitation returns to the atmosphere as vapour through the evaporation process as it falls. Evaporation also occurs from plant, soil and water surfaces. Precipitation that reaches the ground either evaporates, infiltrates into the soil or runs off downstream to ponds, lakes or oceans. Part of soil and surface water is used by plants and animals and returns to the atmosphere through transpiration and respiration. That which percolates through the soil profile seeps into underground streams or reservoirs. The amount of water on the earth is constant; it is always somewhere in the cycle.

Because the total quantity of water available to the earth is finite and indestructible, the global hydrologic cycle is a closed system with any water problem being a distribution (quantity, time, location) or pollution (quality) problem. However, the hydrologic cycle in a river basin, a state like Texas, a county, or a ranch is open. The amount of water received is not a constant. Water disposition within a ranch is influenced by climate, geology and vegetation on the ranch.

The water budget of a ranch depends on precipitation received and the amount of water lost over a given time. The water storage capacity of soil and the losses through evaporation and transpiration (evapotranspiration) influence the change in storage over time. Actual water losses vary depending on the seasonal pattern of rainfall, individual storm intensity and duration, soils, and the kind, amount and distribution of vegetation. Evapotranspiration represents the largest water loss from arid and semiarid lands because runoff and groundwater recharge are relatively minor losses. Evapotranspiration accounts for 90 to 95 per cent of water loss from Texas rangelands. Therefore, efforts to retain as much water as possible on site should theoretically concentrate on reducing evaporation and transpiration losses.

EFFECTS OF VEGETATION ON HYDROLOGIC PROCESSES

The fate of each drop of water falling on the land depends largely on the kind of soil and the vegetative cover. Experience and research have provided land management practices for managing soil and vegetation resources to increase water use efficiency. Adequate vegetation cover prevents erosion by breaking the impact of raindrops and slowing overland flow.

Plant cover reduces soil erosion in raindrop splash by intercepting raindrops and absorbing their energy. Effectiveness of reducing soil splash is proportional to how much cover is present at the time rain occurs. The effective control (95 per cent) of raindrop splash energies requires approximately 2,000 pounds per acre of sodgrass or 3,500 pounds of bunchgrass. Soil-protective values decline rapidly as cover declines below these levels. Plant cover also interrupts the travel of raindrop splash and overland flow thus reducing erosion. Soil movement caused by surface flow depends on the energy of the runoff, the susceptibility of soil to detachment and transportation, and the protection afforded by vegetative cover. Plant cover protects soil from erosive action of runoff water by offering resistance to the movement of water and shielding soil from its effects. Protection from erosion is obtained through resistance of vegetation to the energies of rainfall and runoff.

Generally, however, a combination of plant cover and mechanical measures designed to meet the specific combination of erosion factors operating on a particular land area is necessary for effective erosion control.

Within a particular climate, water loss through transpiration is proportional to the leaf area (transpiring surface) and the availability of water in the rooting zone. The less the transpiring surface and the shallower the root system the less water is lost through transpiration. Soil water content is generally greater under grass cover, due to lower evapotranspiration losses. It is also higher under herbaceous cover than under mixed-brush and herbaceous cover. On the other hand, vegetative cover greatly reduces the amount of runoff with grasses generally decreasing the runoff more than forbs or shrubs.

Therefore, water availability should increase if vegetation conversion is from brush to grass unless some underground geologic layer disrupts normal soil water movement. An impermeable layer at a shallow depth might keep water within reach of the shortest rooting plants. The type of vegetation, because of differences in structure, area and texture of plant surfaces, also influences how much water clings to vegetation and evaporates before passing through the canopy to the ground. Relative interception losses increase from low sodgrasses, to bunchgrasses, to shrubs and trees.

For example, estimated annual interception losses are 10.8 per cent from curly mesquite (sodgrass); 18.1 per cent from sideoats grama (bunchgrass); and 46 per cent from liveoak brush and trees. So, converting from brush or trees to grass should increase the percentage of incoming precipitation available in the soil for use by forage plants. Conversion from brush to grass on rangeland will also theoretically yield more water downstream because infiltration is less under grass than under brush, and runoff is potentially increased if infiltration is reduced. Generally, the amount of cover (biomass), and hence the rate of infiltration, is greatest under trees and shrubs, followed in decreasing order on sites dominated by bunchgrasses, shortgrasses and bare ground. In western Texas grass cover – especially bunchgrasses – provides the most desirable ground cover because it is the most water use efficient. Bunchgrasses also do an excellent job in controlling erosion by holding soil on site and yielding more water off-site than shrubs.

EFFECTS OF VEGETATION MANAGEMENT PRACTICES ON WATER AVAILABILITY

Soil water increases due to vegetation management ultimately depend on whether runoff and deep drainage increase by an amount equal to the reduction in evapotranspiration. Several factors can affect this, including: whether shrub biomass is replaced by grass biomass (if herbaceous cover replaces shrub cover in equal amounts there will be little difference in transpiration), speed of percolation of water in the soil profile (restrictive layers may slow water percolation and allow more transpiration), high rainfall areas that get more water than replacement plants use, whether transpiring tissue of grasses is less than the trees or shrubs it replaces, storm characteristics. Vegetative surfaces can hold only a certain amount of water at a given time. Large storms account for the major portion of runoff and deep drainage in the Southwest.

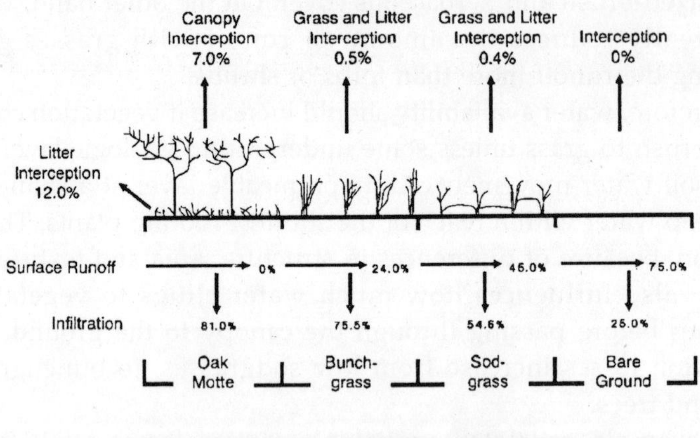

Fig:2. Vegetation type Greatly Influences what Happens to Incoming Precipitation

Vegetation management practices can affect both on-site and off-site water through their effects on vegetation composition and soil surface characteristics. The amount and quality of increased soil water depend on the original vegetation, soils and climate. Also it depends on the range management practice used in conversion of the vegetation and its effects on vegetative composition and soil surface characteristics. Any practice that increases standing vegetation and litter will decrease runoff and sediment production.

VEGETATION MANAGEMENT PRACTICES

Range management practices directly affecting vegetation are:

- Grazing management,
- Range revegetation.
- Brush and weed management.

Grazing Management

Ability to control kinds and numbers of animals and when they utilise the rangeland is absolutely essential in regulating the effects of grazing on vegetation. Of all the range management practices and technologies available, proper stocking or control of forage utilisation is most important. Continued excessive defoliation is the major cause of range deterioration. Deteriorated range means more runoff and erosion. Without control of animal numbers, the season of use, and distribution of animals, other practices are usually of limited value in maintaining desirable vegetation cover. Any grazing management strategy that enhances vegetative cover improves water use efficiency and conserves the soil resource. Grazing management should be the first consideration in developing water management strategies.

Range Revegetation

Artificial revegetation utilises agronomic practices to restore native plant communities or to introduce desired species. It is an expensive and ecologically disruptive process, particularly risky in arid and semi-arid areas. Artifical revegetation should not be attempted unless natural revegetation through grazing management will not restore the range to the desired condition within an acceptable period.

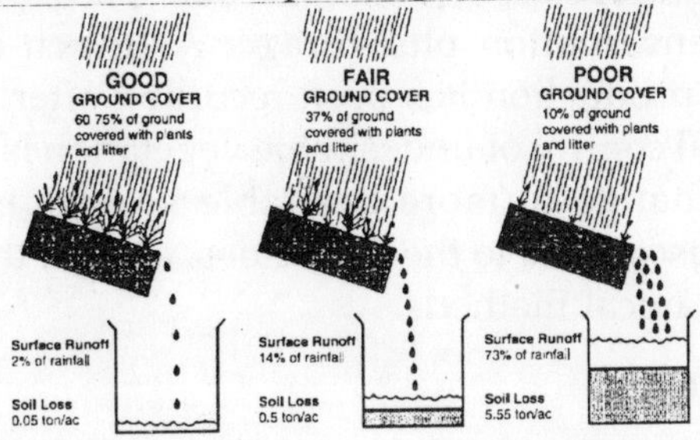

Fig:3. Good Ground Cover Means Less Runoff and Erosion and Better Water Use Effuciency

In general, artificial revegetation is not recommended if 10 per cent or more of the vegetative cover is made up of desirable species. Revegetation may also not be feasible because of poor soil conditions, erosion hazard or economic considerations.

Seeding in conjunction with land surface modification techniques such as water spreading, water harvesting, contour furrowing, pitting and diking will enhance the probability of success for reseeding and managing water use. Seeding should be considered on severely depleted ranges and where vegetation modification practices call for a change in species. It is a valuable practice in replacing one plant cover with another to provide the desired forage or water management.

Brush and Weed Management

Development of ecologically sound land management practices requires a clear understanding of how the practices affect the hydrology of the site. Reducing the density of undesirable species increases the availability of moisture and nutrients for more desirable forage species. Also, water yields should increase if brush removal reduces transpiration losses. Control methods include burning, mechanical control, herbicidal control and biological control. The hydrological impact of all the methods has not been studied specifically, but many studies allude to increased on-site moisture availability. Others report increased downstream flow. Each method affects the water regime differently.

Herbicidal Control

Defoliation of plants with herbicides immediately affects evapotranspiration losses. When plants defoliate, transpiration is reduced and litter is added to the soil increasing soil aggregation and water infiltration. Dead stems and roots decay and leave organic matter on the surface and in holes created by decomposing roots. The total impact is that both the on-site and off-site water regime is enhanced without risk of erosion or increase in sediment loads due to physical disturbance. Replacement of brush species with herbaceous species, especially grasses, provides water conservation plus forage. As much as a 500 per cent increase in forage production has been recorded after herbicidal control of brush. Herbicidal control of undesirable vegetation is applicable where: (1) vegetational change to more desirable species is needed, (2) the undesirables are susceptible to the herbicides, and (3) the terrain does not lend itself to mechanical methods.

Mechanical Control

Mechanical methods such as axing, shredding or roller chopping add litter to soil with relatively little soil disturbance. These methods are applicable on nonsprouting species or where retreatments can be applied. Dozing, root-plowing, grubbing or chaining remove plants from the soil and create considerable soil disturbance. Bulldozing and grubbing create pits where trees and roots are extracted.

These pits act as water catchments which concentrate water nutrients and enhance moisture infiltration. Evapotranspiration is reduced making more water available for replacement plant use or deep percolation. However, the reduction will not be maintained unless another vegetative cover is established. Erosion is a hazard until herbaceous vegetation is re-established. These methods are applicable on most nonrocky soils and where vegetative cover can be replaced. Chaining, cabling or dragging usually does not increase runoff and erosion if debris and litter are left in place to protect the soil and the herbaceous vegetation is re-established. Generally, soil moisture and runoff are much higher on chained areas than unchained areas throughout the year.

These differences are due to changes in the microclimate, mulching effect of the litter and differences in water accumulation. Innovations that make chaining even more effective include the disk chain and diskchain-diker. Chaining and dragging are applicable on large acreages and work best with moist soil conditions and single stem non-sprouting species. Chaining, dragging and cabling should be considered temporary and followed in due time with repeat applications or other follow-up methods of control. The relative impact of any plant control technique on the water regime depends on several factors:

- Severity of the soil disturbance.
- The response of the herbaceous vegetative.
- Effectiveness of the control method in removing brush.
- The impact of the practice on litter and ground cover.
- Time since implementation of the practice.

SOIL MODIFICATION PRACTICES

Soil modification practices used for range improvement generally bring about improvements in range productivity through increased conservation of water and water use efficiency. Many farming techniques and implements have been adapted for range use, and some have been developed specifically for range use.

Contour Furrowing

Contour furrows are grooves or ditches made in soil by various implements (plows, chisels, furrowers, etc.). The furrows should be placed on the contour to collect runoff water and increase soil water storage. Furrow dimensions vary considerably from grooves 4 to 6 inches wide and 3 to 4 inches deep to as much as 2 feet wide and a foot deep. Interrupting the furrows with dams at intervals increases their effectiveness in ponding water and increasing infiltration.

Contour furrows have been used successfully in semi-arid regions to reduce runoff and improve infiltration for increased forage production. Contour furrowing of poor condition range has been shown to reduce runoff and conserve more than one inch of water annually in the Great Plains. Increased herbage production follows improved water retention and storage and the transport of nutrients from surface layers to lower depths.

Contour furrowing is applicable on productive soils of restricted permeability on long uniform slopes with simple contour patterns. They are most effective when rainfall intensities do not greatly exceed the hydraulic properties of the furrows.

Terracing

Terraces differ from furrows in that they are larger and applied on the grade to allow controlled runoff. They are designed primarily for flood control and reduction of runoff and sedimentation on moderately steep slopes. Although terraces have been widely used in restoring critical watersheds in the West, their use is generally impractical except as a watershed treatment practice on rangeland.

Pitting

The creation of small basins or pits to catch and hold precipitation on the site has been used since the dust bowl days of the 1930s. Known as pitting, it is often used in conjunction with reseeding to enhance seedling establishment by concentrating nutrients and water. Tools used for pitting vary widely. Almost any equipment capable of gouging, digging or in some way creating pits in the soil surface can be used. The most commonly used implements are:

- Modifications of diskplows.
- Spike-toothed pitters.

Modified disk-plows gouge out long shallow pits while the spike type pitter creates small basins. Modifications of spike-tooth pitters are called aerators. Aerators use spikes or cleats to create pits and aerate the soil increasing water and air movement. Pitting has been effective in increasing forage production by as much as 100 per cent, primarily due to enhanced water relations.

The disturbance and better water relations increase productivity of the remaining vegetation and, through plant succession, make better plant communities. The value of pits in water retention depends on their density, size, depth and soil permeability. The pit effectively serves as a basin to collect water and allow soil penetration. Pitting is best suited to medium textured soils with less than 8 per cent slope. Its value is limited on sandy, rocky or brush covered soils.

Ripping

Ripping on rangeland is synonymous with chiseling or subsoiling on farmland. It is done to fracture compacted soil layers to allow water and root penetration.

Implements used include chisel-type plows capable of penetrating to depths below soil hardpans to 36-inch depths. However, 12 inches or less is the most common depth of ripping on rangeland. Because of the soil disturbance and furrowing effect, ripping increases soil water penetration and can dramatically increase forage production. Ripping can reduce surface runoff dramatically.

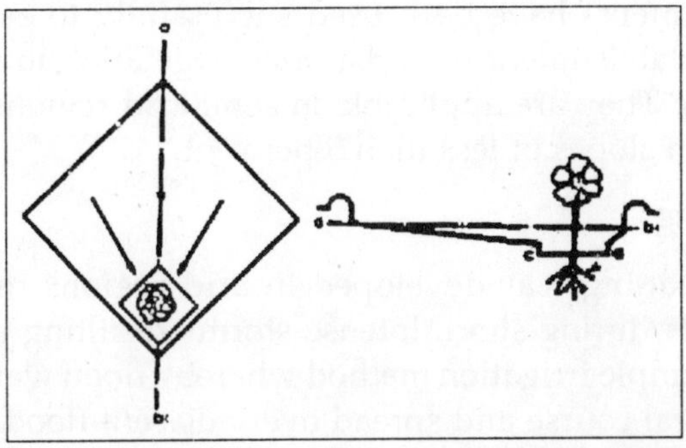

Fig:4. Chiseling or Aerating Coastal Bermudagrass, Kleingrass and Suffelgrass may Increase Forage Yield by 100 to 300 per cent

It is applicable on medium to fine textured soils with compacted soil layers. Ripping should be done on the contour, and under dry soil conditions. Forage yield increases of 100 to 300 per cent have been obtained from chiseling or aerating coastal bermudagrass, kleingrass and buffelgrass in south Texas.

WATER HARVESTING PRACTICES

Microcatchments

Collecting runoff water by creating small basins or microcatchments is practiced in arid and semi-arid regions throughout the world. The goal of microcatchments is to catch water and allow its storage in soil rather than runoff.

Fig:5. Schematic of a Microcatchment. Microcatchments Consist of a Catchment Area an Infiltration Basin

The water stored can be used for plant growth or ground water recharge. The basins concentrate water in a small area and provide extra water for plant establishment. Once plants are established, the catchment continues to collect water and nutrients for plant growth. The water that can be stored depends upon the size of the microcatchment and ultimately the effective depth of soil profile.

Microcatchments have been used successfully to establish saltbush and enhance establishment of herbaceous vegetation in the Trans Pecos region of Texas. They are applicable in semi-arid regions on medium to fine soils and on slopes of less than 5 per cent.

Water Spreading

Water spreading was developed in arid regions receiving limited rainfall that falls during short, intense storms resulting in runoff. Water spreading is a simple irrigation method whereby flood waters are diverted from their natural course and spread over adjacent flood plains. Ditches, dikes, small dams, rock, brush and wire fences are used to divert flood flows and spread the water over the flood plain to allow infiltration. The water that penetrates is then available for plant growth or deep percolation.

The earth's water cycle is a closed system with a finite but indestructible quantity. The water budget of a particular ranch, however, is open and depends upon how much precipitation is received and how much water leaves the ranch. The kind, amount and distribution of vegetation are the major variables affecting water use and loss from a range site. Within a particular climate, water loss through transpiration is proportional to the leaf area and the availability of water in the rooting zone. Grasses, especially bunchgrasses, provide the most desirable ground cover because they are the most water efficient, control erosion very well, and yield more runoff water off-site than shrubs.

Soil water content is greater under herbaceous cover, especially grasses, than under mixed-brush and herbaceous cover. Water yield increases resulting from vegetation management ultimately depend upon whether runoff and deep drainage water losses exceed transpiration losses. Range management practices can affect both on-site and off-site water through their effects on vegetation composition and soil characteristics. The amount and quality of increased water availability depend on the original vegetation, soils and climate. Range improvement practices such as grazing management, revegetation, brush and weed management, and soil modification techniques directly affect the water regime.

Proper stocking should be the first consideration in range water management. Any grazing management strategy that enhances vegetative cover improves the water regime both in the pasture and downstream. Range seeding should be considered on severely depleted range and where vegetation modification practices call for a change in species. Seeding in conjunction with land surface modification techniques will enhance the probability of success.

Brush control with herbicides reduces transpiration losses, adds litter to the soil surface and increases water infiltration. Mechanical brush control methods enhance the water regime by creating soil disturbance, reducing transpiration and increasing infiltration. Both the on-site and off-site water regimes are enhanced. Soil modification practices generally bring about increases in range productivity through increased conservation of water and improved water use efficiency. Practices applicable to Texas rangelands include contour furrowing, pitting, ripping, microcatchments and water spreading. Water harvesting practices such as microcatchment and water spreading systems collect runoff water that would otherwise be lost.

VEGETATION

Vegetation is a general term for the plant life of a region; it refers to the ground cover life forms, structure, spatial extent or any other specific botanical or geographic characteristics. It is broader than the term *flora* which refers exclusively to species composition. Perhaps the closest synonym is plant community, but *vegetation* can, and often does, refer to a wider range of spatial scales. Primeval redwood forests, coastal mangrove stands, sphagnum bogs, desert soil crusts, roadside weed patches, wheat fields, cultivated gardens and lawns; are all encompassed by the term *vegetation*.

Vegetation supports critical functions in the biosphere, at all possible spatial scales. First, vegetation regulates the flow of numerous biogeochemical cycles, most critically those of water, carbon, and nitrogen; it is also of great importance in local and global energy balances. Such cycles are important not only for global patterns of vegetation but also for those of climate.

Second, vegetation strongly affects soil characteristics, including soil volume, chemistry and texture, which feed back to affect various vegetational characteristics, including productivity and structure. Third, vegetation serves as wildlife habitat and the energy source for the vast array of animal species on the planet (and, ultimately, to those that feed on these).

Vegetation is also critically important to the world economy, particularly in the use of fossil fuels as an energy source, but also in the global production of food, wood, fuel and other materials. Perhaps most importantly, and often overlooked, global vegetation (including algal communities) has been the primary source of oxygen in the atmosphere, enabling the aerobic metabolism systems to evolve and persist. Lastly, vegetation is psychologically important to humans, who evolved in direct contact with, and dependence on, vegetation, for food, shelter, and medicine.

Classification

Biomes Classified by Vegetation

Much of the work on vegetation classification comes from European and North American ecologists, and they have fundamentally different approaches. In North America, vegetation types are based on a combination of the following criteria: climate pattern, plant habit, phenology and/or growth form, and dominant species. In the current US standard (adopted by the Federal Geographic Data Committee (FGDC), and originally developed by UNESCO and The Nature Conservancy), the classification is hierarchical and incorporates the non-floristic criteria into the upper (most general) five levels and limited floristic criteria only into the lower (most specific) two levels. In Europe, classification often relies much more heavily, sometimes entirely, on floristic (species) composition alone, without explicit reference to climate, phenology or growth forms. It often emphasises indicator or diagnostic species which separate one type from another.

In the FGDC standard, the hierarchy levels, from most general to most specific, are: *system, class, subclass, group, formation, alliance,* and *association*. The lowest level, or association, is thus the most precisely defined, and incorporates the names of the dominant one to three (usually two) species of the type.

An example of a vegetation type defined at the level of class might be "*Forest, canopy cover > 60%*"; at the level of a formation as "*Winter-rain, broad-leaved, evergreen, sclerophyllous, closed-canopy forest*"; at the level of alliance as "*Arbutus menziesii* forest"; and at the level of association as "*Arbutus menziesii-Lithocarpus densiflora* forest".

Structure

A primary characteristic of vegetation is its three-dimensional structure, sometimes referred to as its *physiognomy*, or architecture. Most people have an understanding of this idea through their familiarity with terms like "jungle", "woods", "prairie" or "meadow"; these terms conjure up a mental image of what such vegetation looks like. So, meadows are grassy and open, tropical rainforests are dense, tall and dark, savannas have trees dotting a grass-covered landscape, etc.

Obviously, a forest has a very different structure than a desert or a backyard lawn. Vegetation ecologists discriminate structure at much more detailed levels than this, but the principle is the same. Thus, different types of forests can have very different structures; tropical rainforests are very different from boreal conifer forests, both of which differ from temperate deciduous forests. Native grasslands in South Dakota, Arizona, and

Indiana are visibly different from each other, low elevation chaparral differs from that at high elevations, etc. Structure is determined by an interacting combination of environmental and historical factors, and species composition.

It is characterized primarily by the horizontal and vertical distributions of plant biomass, particularly foliage biomass. Horizontal distributions refer to the pattern of spacing of plant stems on the ground. Plants can be very uniformly spaced, as in a tree plantation, or very non-uniformly spaced, as in many forests in rocky, mountainous terrain, where areas of high and low tree density alternate depending on the spatial pattern of soil and climatic variables.

Three broad categories of spacing are recognised: uniform, random and clumped. These correspond directly to the expected variation in the distance between randomly chosen locations and the closest plant to such locations. Vertical distributions of biomass are determined by the inherent productivity of an area, the height potential of the dominant species, and the presence/absence of shade tolerant species in the flora. Communities with high productivities and in which at least one shade tolerant tree species is present, have high levels of biomass because of their high foliage densities throughout a large vertical distance.

Ecologists thus often measure a surrogate, plant *cover*, which is defined as the percentage of the ground surface area that has plant biomass (especially foliage) vertically above it. If the vertical distribution of the foliage is broken into defined height layers, cover can be estimated for each layer, and the total cover value can therefore be over 100; otherwise the values range from zero to 100. The measure is designed to be a rough, but useful, approximation of biomass.

In some vegetation types, the underground distribution of biomass can also discriminate different types. Thus a sod-forming grassland has a more continuous and connected root system, while a bunchgrass community's is much less so, with more open spaces between plants (though often not as drastic as the openings or spacings in the above-ground part of the community, since root systems are generally less constrained in their horizontal growth patterns than are shoots). However, below-ground architecture is so much more time-consuming to measure, that vegetation structure is almost always described in relationship to the above-ground parts of them community.

DYNAMICS

Like all biological systems, plant communities are temporally and spatially dynamic; they change at all possible scales. Dynamism in vegetation is defined primarily as changes in species composition and/or vegetation structure.

Temporal Dynamics

Temporally, a large number of processes or events can cause change, but for sake of simplicity they can be categorised roughly as either abrupt or gradual. Abrupt changes are generally referred to as disturbances; these include things like wildfires, high winds, landslides, floods, avalanches and the like. Their causes are usually external (exogenous) to the community—they are natural processes occurring (mostly) independently of the natural processes of the community (such as germination, growth, death, etc).

Such events can change vegetation structure and species composition very quickly and for long time periods, and they can do so over large areas. Very few ecosystems are without some type of disturbance as a regular and recurring part of the long term system dynamic. Fire and wind disturbances are particularly common throughout many vegetation types worldwide. Fire is particularly potent because of its ability to destroy not only living plants, but also the spores and seeds representing the potential next generation, and because of fire's impact on faunal populations and soil characteristics.

Temporal change at a slower pace is ubiquitous; it comprises the field of ecological succession. Succession is the relatively gradual change in structure and composition that arises as the vegetation itself modifies various environmental variables, including light, water and nutrient levels over time.

These modifications change the suite of species most adapted to grow, survive and reproduce in an area, causing floristic changes. These floristic changes contribute to structural changes that are already inherent in plant growth even in the absence of species changes (especially where plants have a large maximum size, i.e., trees), causing slow and broadly predictable changes in the vegetation. Succession can be interrupted at any time by disturbance, setting the system either back to a previous state, or off on another trajectory altogether. Because of this, successional processes may or may not lead to some static, final state. Moreover, accurately predicting the characteristics of such a state, even if it does arise, is not always possible. In short, vegetative communities are subject to many and unpredictable variables that limit predictability.

Spatial Dynamics

As a general rule, the larger an area under consideration, the more likely the vegetation will be heterogeneous across it. Two main factors are at work. First, the temporal dynamics of disturbance and succession are increasingly unlikely to be in synchrony across any area as the size of that area increases. That is, different areas will be at different developmental stages due to different local histories, particularly their times since last major disturbance.

This fact interacts with inherent environmental variability, which is also a function of area. Environmental variability constrains the suite of species that can occupy a given area, and the two factors together interact to create a mosaic of vegetation conditions across the landscape. Only in agricultural or horticultural systems does vegetation ever approach perfect uniformity. In natural systems, there is always heterogeneity, although its scale and intensity will vary widely. A natural grassland may seem relatively homogeneous when compared to the same area of partially burned forest, but highly diverse and heterogeneous when compared to the wheat field next to it.

GLOBAL VEGETATION PATTERNS AND DETERMINANTS

At regional and global scales there is predictability of certain vegetation characteristics, especially physiognomic ones, which are related to the predictability in certain environmental characteristics. Much of the variation in these global patterns is directly explainable by corresponding patterns of temperature and precipitation (sometimes referred to as the energy and moisture balances). These two factors are highly interactive in their effect on plant growth, and their relationship to each other throughout the year is critical. Such relationships are shown graphically in climate diagrams.

By graphing the long term monthly averages of the two variables against each other, an idea is given as to whether or not precipitation occurs during the warm season, when it is most useful, and consequently the type of vegetation to be expected. For example, two locations may have the same average annual precipitation and temperature, but if the relative timing of the precipitation and seasonal warmth are very different, so will their vegetation structure and growth and development processes be.

Vegetation scientists study the causes of the patterns and processes observed in vegetation at various scales of space and time. Of particular interest and importance are questions of the relative roles of climate, soil, topography, and history on vegetation characteristics, including both species composition and structure. Such questions are often large scale, and so cannot easily be addressed by experimentation in a meaningful way. Observational studies supplemented by knowledge of botany, paleobotany, ecology, soil science etc, are thus the rule in vegetation science.

Pre-1900

Vegetation science has its origins in the work of botanists and/or naturalists of the 18th century, or earlier in some cases. Many of these were world travellers on exploratory voyages in the Age of Exploration, and their work was a synthetic combination of botany and geography that today we would call plant biogeography (or *phytogeography*). Little was known about worldwide floristic or vegetation patterns at the time, and almost nothing about what determined them, so much of the work involved collecting, categorising, and naming plant specimens. Little or no theoretical work occurred until the 19th century.

The most productive of the early naturalists was Alexander von Humboldt, who collected 60,000 plant specimens on a five year voyage to South and Central America from 1799 to 1804. Humboldt was one of the first to document the correspondence between climate and vegetation patterns, in his massive, life-long work "Voyage to the Equinoctial Regions of the New Continent", which he wrote with Aimé Bonpland, the botanist who accompanied him. Humboldt also described vegetation in physiogonmic terms rather than just taxonomically.

His work presaged intensive work on environment-vegetation relationships that continues to this day. The beginnings of vegetation study as we know it today began in Europe and Russia in the late 19th century, particularly under Jozef Paczoski, a Pole, and Leonty Ramensky, a Russian. Together they were much ahead of their time, introducing or elaborating on almost all topics germane to the field today, well before they were so in the West. These topics included plant community analysis, or phytosociology, gradient analysis, succession, and topics in plant ecophysiology and functional ecology. Due to language and/or political reasons, much of their work was unknown to much of the world, especially the English-speaking world, until well into the 20th century.

Post-1900

In the United States, Henry Cowles and Frederic Clements developed ideas of plant succession in the early 1900s. Clements is famous for his now discredited view of the plant community as a "superorganism". He argued that, just as all organ systems in an individual must work together for the body to function well, and which develop in concert with each other as the individual matures, so the individual species in a plant community also develop and cooperate in a very tightly coordinated and synergistic way, pushing the plant community towards a defined and predictable end state.

In contrast to Clements, several ecologists have since demonstrated the validity of the individualistic hypothesis, which asserts that plant communities are simply the sum of a suite of species reacting individually to the environment, and co-occurring in time and space. Ramensky initiated this idea in Russia, and in 1926, Henry Gleason developed it in a paper in the United States.

Gleason's ideas were categorically rejected for many years, so powerful was the influence of Clementsian ideas. However, in the 1950s and 60s, a series of well-designed studies by Robert Whittaker provided strong evidence for Gleason's arguments, and against those of Clements. Whittaker, considered one of the brightest and most productive of American plant ecologists, was a developer and proponent of gradient analysis, in which the abundances of individual species are measured against quantifiable environmental variables or their well-correlated surrogates.

In studies in three very different montane ecosystems, Whittaker demonstrated strongly that species respond primarily to the environment, and not necessarily in any coordination with other, co-occurring species. Other work, particularly in paleobotany, has lent support to this view at larger temporal and spatial scales.

Recent Developments

Since the 1960s, much research into vegetation has revolved around topics in functional ecology. In a functional framework, taxonomic botany is relatively less important; investigations centre around morphological, anatomical and physiological classifications of species, with the aim of predicting how particular groups thereof will respond to various environmental variables.

The underlying basis for this approach is the observation that, due to convergent evolution and (conversely) adaptive radiation, there is often not a strong relationship between phylogenetic relatedness and environmental adaptations, especially at higher levels of the phylogenetic taxonomy, and at large spatial scales. Functional classifications arguably began in the 1930s with Raunkiaer's division of plants into groups based on the location of their apical meristems (buds) relative to the ground surface.

Functional classifications are crucial in modeling vegetation-environment interactions, which has been a leading topic in vegetation ecology for the last 30 or more years. Currently, there is a strong drive to model local, regional and global vegetation changes in response to global climate change, particularly changes in temperature, precipitation and disturbance regimes.

Functional classifications such as the examples above, which attempt to categorise all plant species into a very small number of groups, are unlikely to be effective for the wide variety of different modelling purposes that exist or will exist. It is generally recognised that simple, all-purpose classifications will likely have to be replaced with more detailed and function-specific classifications for the modeling purpose at hand. This will require much better understanding of the physiology, anatomy, and developmental biology than currently exists, for a great number of species, even if only the dominant species in most vegetation types are considered.

VEGETATION, SOILS & WATER EFFECTS OF RECREATION ON ROCKY MOUNTAIN WILDLIFE

Vegetation, soil, and water, structurally arranged to meet the life-cycle necessities of wildlife, *is* wildlife habitat. The healthy function and interaction of these ecological components largely determines the well-being of wildlife populations. Soils, and water is provided to initiate dialogue on this topic and to invite a more comprehensive synthesis of the extensive body of literature that exists relative to recreational impacts upon vegetation, soils, and water.

Disturbances to landscapes induced by recreational activities were documented as early as 1935 by Bates who described mechanical damage to plants and changes in soil structure. Recreational impacts to vegetation can be extensive because recreationists disperse throughout large areas although the most visible impacts are generally near denuded trails, campsites and animal holding facilities. Intensity of recreational use in some areas has resulted in water quality degradation potentially affecting human health. Activities in the name of leisure certainly have the ability to impact wildlife habitat, and possibly the biosphere. As Odum points out, "We need to learn a lot more about how the current real-world life-support systems of our earth function. so we can maintain the quality of these systems."

Damage to Vegetati ›n from Trampling

Vegetation is affected by trampling, which initially bends and weakens leaves and branches and ultimately breaks them. Trampling directly damages plants reducing photosynthetic surfaces, seed production, and carbohydrate reserves. It eventually kills some vegetation species. This changes community composition. As soil compacts and erodes, roots are exposed and eventually killed. This level of disturbance allows the establishment of weed species. The greatest increase in plant mortality from trampling on simulated trails occurred at a low intensity of trampling, that is, between 100 and 300 passes on the simulated trail, with gradual increases in mortality at higher intensities. Alpine vegetation is very sensitive to trampling. Significance of impact is related to rarity of the vegetation type, the value of the vegetation to wildlife species that depend on it, and to the potential to destabilise soils. Trails along streams negatively affect riparian vegetation with concurrent increases in sedimentation to adjacent streams. Sediments can inhibit or kill periphyton communities, bacteria, and fungi, which are important food sources for invertebrates, amphibians, and fish.

Damage to Vegetation from Snowmobiles and OHVs

Snowmobile impacts appear to be greater in forest communities than in open areas, partially because snow drifting fills in the tracks in open areas. One traverse over undisturbed snow can affect the physical environment beneath the snow and physically damage important plants. Snow compaction lowers soil temperatures and reduces the survival of plants and soil microbes. The impact of snowmobiling on the biota varies with the depth of snow accumulation, the intensity of snowmobile traffic, and the susceptibility of the organism to injury caused by cold temperatures or physical contact. Temperatures beneath snow compacted by snowmobiles are considerably colder than those under undisturbed snow cover.

Thus the growth and reproductive success of early spring flowers is retarded and reduced where snowmobiles travel. Herbs with large underground storage organs are winter-killed under snowmobile tracks. Woody plants are particularly vulnerable to physical damage by snowmobiles. "The greater the torque applied at the machine/environment interface, the greater the potential for impact." Boucher and Tattar reported that disruption to plants under snowmobile trails was most heavy on south-facing slopes exceeding 30 degrees incline. On steep slopes, shallow roots and rhizomes were damaged. Neumann and Merriam also showed that direct mechanical effects by snowmobiles on vegetation at and above snow surface can be severe. After only a single pass by a snowmobile, more than 78% of the saplings on the trail were damaged, and nearly 27% of them were damaged seriously enough to cause a high probability of death.

Although Neumann and Merriam indicated that it is difficult to predict whether whole ecosystems will be affected by snowmobiling, they state that "management of snowmobiles as a factor in the human environment should not await this purely ecological information". Vegetation suffers directly and indirectly from the passage of off-road vehicles (ORVs). The effects can last decades or even centuries:

> "There is a strong correlation between damage to soil and damage to vegetation. Compaction and erosion, for instance, influence the ability of plants to take up nutrients and carbon dioxide, experience proper root growth, and have enough stability to grow upwards. Unless regulations exist and are strictly enforced, users will choose their own routes and hillclimb areas. Unfortunately, they select areas for their challenge, not for their soil type and stability."

A controlled study by Leininger and Payne showed that forbs were damaged by all-terrain vehicles (ATVs) most significantly in early fall. Shrubby species were impacted most during spring and early summer.

Graminoids were least affected from vehicle travel. Eight passes with a vehicle caused significant loss of shrub cover. Vegetation was completely destroyed after passes with the vehicle. In a study of the impact of ORV use on valley vegetation in West Virginia, Stout indicated that valley grasses tend to replace the shrubs that have been destroyed:

> "Direct and indirect impacts add up to a transformation of plant communities for the worse. Communities on a site prior to human disruption are composed of a variety of species adapted to that particular habitat. They live together in a relatively stable balance. ORVs and other severe disruptions destroy the balance and make it impossible for the plants to continue to coexist. Some plants are better able to endure the presence of ORVs than others. The adaptable plants are likely to have root systems that survive compaction. These plants flourish while more sensitive species disappear."

Recreational Activities and Weeds

Uncontrolled and largely unmanaged trail systems, spreading across public lands, provide an expeditious avenue for weed dispersal. Recreational activities may introduce and encourage weeds. Hay for pack animals and the resulting excrement are sources of weed seeds. Backpackers may import seeds on their equipment, and motorised vehicles are capable of distributing weed seeds.

Rapid dispersal of weeds is characteristic of motorised routes where a vehicle in one trip, can spread 2,000 knapweed seeds over a 10-mile course. Although dispersal of weeds is a prodigious issue; disturbance of soils by vehicles has long-term effects that favour the establishment of weedy species.

After disturbance, weedy, often exotic species, are likely to gradually crowd out native vegetation; thus biodiversity may be drastically lowered. Given establishment, the current political climate demands that we kill weeds with potent herbicides. In Montana all land management agencies are operating under state legislative mandates that weeds be "controlled". Thus the war on weeds is being conducted by an army of personnel with backpack sprayers, tanks on ATVs, larger tanks mounted on trucks with boom-mounted applicators, as well as aerial spraying. Tons of diverse herbicides are infused into the environment. One hypothesises that this generalised approach will remove all forbs, and allow noxious forbs known to be especially invasive, to re-invade with greater force. Two studies support this idea.

Long-term (20 year) observations of a fescue-grassland sprayed for spurge, show a good initial kill of all forbs, a strong reinvasion of spurge, but no reinvasion of native forbs. Long-term (30 year) observations of *Abies lasiocarpa/Vaccinium scoparium* habitat types sprayed to minimise understory development and encourage tree establishment, have some herbs and grasses in the understory, but no shrubs. Further testing is critically important before widespread non-directed spraying can be justified. Such testing could be made by analyzing previously sprayed sites or requiring applicator follow-up to monitor incidental effects on native plants by their treatments.

Erosion, Compaction and Rutting of Soils

Soil impacts caused by recreation depend on soil type, vegetation cover, topography, and intensity of use. Recreational impacts to soils are found at picnic areas and campgrounds, low-standard roads, trails, and off-road. The primary detrimental soil impacts are loss of productivity, erosion, compaction, rutting, and displacement.

Soils derived from parent materials such as granitics, coarse-grained volcanics, or gneisses are sandy and easily erode when organic layers or vegetative cover is removed. This condition is made worse by moderately steep or steeper slopes and south to west aspects. Erodibility describes a soil's susceptibility to erosion and is influenced by properties such as texture, structure, organic matter content, and chemical make-up. Erosion can be so severe that soil-surface horizons are lost, and long-term soil productivity is decreased.

The process of physically detaching and transporting soil particles, can be further accelerated by a decrease in infiltration capacity associated with horse and off-highway vehicle (OHV, includes full-sized vehicles as well as ATVs) use. Off-road vehicles exert shear stress parallel to the soil surface, as well as compressive stress. Direct effects of sheer and compression stress include crushing of foliage, root systems, and seedlings, and uprooting and disruption of root systems of larger plants by shear stresses to the soil.

Indirect effects include undercutting of root systems as vehicle paths are enlarged by erosion, creation of new erosion channels on land adjacent to vehicle-destabilized areas due to accelerated runoff or wind erosion, burial by debris eroded from areas used by vehicles, and reduction of biological capability of the soil by physical modification and stripping of the more fertile upper soil layers.

There are two basic responses in soils from OHV use: "One, sandy and gravelly soils are susceptible to direct quarrying by ORVs, and when stripped of vegetation they are susceptible to rapid erosion processes – usually by rill and gully erosion. Two, more clay-rich soils are less sensitive to direct mechanical displacement by ORVs, but the rates of erosion of stripped clay-rich soil are much higher under ORV use than under natural conditions." Also, pounding of the latter creates strong seal on the surface, which reduce water infiltration. This causes gullying lower in the drainage due to greater rainwater runoff.

As with agricultural soils, conservation of soils in ORV areas requires intensive on-site management. The standards of tolerable soil loss (in tons/acre/year) set forth by the SCS are predicated on retention of a long-term designated level of biological productivity of the land. Protection of public lands used by ORVs may be accomplished under the same concept of loss tolerances devised from long experience and research with agricultural uses of soil.

Application of such standards, of course, requires a thorough description of the soil resources of the areas of intended use and intensive management to control and monitor the levels of deterioration so that the designated level of permissible loss is not exceeded. The problems of rapid deterioration of vegetation and soil are most severe in the upland and arid areas most commonly selected for vehicle use in the west.

Shallow water tables associated with riparian areas are especially susceptible to detrimental rutting, erosion, displacement, and compaction when they are crossed by low-standard roads, trails, or used for off-road travel by horses, mountain bikes, motorcycles, ATVs, and full size vehicles. Intense recreational use of these areas can lead to increased streambank erosion, bank failure, and loss of wetland function including water storage and sediment trapping. Soils in semi-arid and arid areas are also impacted by off-road vehicles, mountain bikes, and foot traffic.

Once the soil surface and microbiotic crusts are disturbed, erosion by wind, rain and gravity increase. Soil compaction from even moderate traffic increases runoff and soil displacement, causing concentrations of water that make erosive forces more effective. Reestablishment of crusts may require hundreds of years.

These levels of compaction reduce water infiltration, increase runoff, limit the ability of vegetation to reestablish, and result in an overall loss of soil productivity. Vegetation recovers little 20 years after one summer of trampling in an *Abies/Vaccinium scoparium* habitat type. Wanek's research indicated that bacterial decomposers and litter decomposition may be significantly affected by roads. Indicates that a four-wheel drive vehicle with a track 47 cm wide will impact at least a hectare of ground for every 23 km traveled.

In contrast, the average motorcycle track is 13 cm wide, and impacts a hectare every 77 km. Mass failures, such as landslides, occur naturally, but roads dramatically increase their frequency and magnitude, from several times to hundreds of times. Such failures can be a major sediment source detrimental to insects and fish of nearby stream.

Water

Water Quality and Quantity Issues Related to Roads

Roads have been identified as the major impact on the forest environment. Natural drainage patterns can be disrupted by roads when water is diverted and prevented from infiltrating into soils. Impacts from roads basically fall into three areas: introduced sediment into streams; snowmelt re-direction and concentration, and surface flow production.

Roads can affect both the volume of water available as surface runoff and the efficiency by which water flows through a watershed. Roads accelerate water flows and sediment transport, which raise flood levels and degrade aquatic ecosystems. Local hydrologic and erosional effects along roads are dispersed across the land, whereas major impacts are concentrated in the stream network and distant valleys. "Roads increase the volume of surface runoff in two ways. First, compacted road surfaces do not readily absorb water. And, second, road cuts intercept subsurface water flow and convert it to surface flow. Water moving through a watershed as surface runoff moves more quickly because it has less resistance to flow compared to water percolating through soil. The faster surface runoff causes accelerated soil erosion. Roads thus increase water reaching stream channels during a storm and snowmelt events, so channels must accommodate the additional volume of water and road-related sediment. More water and sediment in channels alter their physical structure, usually with negative effects on aquatic habitat."

Aquatic ecosystems are impacted by sediments from roads as well as the large water flushes just described. Road surface drainage, and the sagging of road ditches into channels and creeks, is a common Best Management Practice (BMP) violation, and the dwindling national forest road maintenance budget makes it difficult to maintain culvert crossings. Sediment originating from roads reach streams and rivers, degrading habitat and impairing fish reproduction. Fine sediments impact spawning habitat by settling into and covering spawning gravels, and interfering with salmonid reddnest construction. Excessive sediments can impede intergravel water flow that provides oxygen and removes waste products, both of which are necessary for successful egg development. Roads thus increase barriers to migrating adult and juvenile salmonids and the macroinvertebrates. When culverts fail during storm and runoff events, tremendous amounts of sediment can be delivered directly to the channel and from there down into lower streams, potentially affecting sensitive fish habitat.

"Even on roads that appear to be so thick with alder that a sediment production concern seems ludicrous, we often find that the road tracks are still actively functioning as erosion sources." Spring snowmelt and runoff from frequent mid-winter melt and rain-on-snow events that would normally travel in a downhill direction, usually as shallow sub-surface flow, is intercepted by the compacted roads and their ditches and becomes surface flow. Thus the drainage efficiency of a watershed is dramatically increased. In a study of hydrologic interaction of forest roads with stream networks, it was pointed out that the contributions of the road and ditch network provides a substantial peak in water runoff flows, and may extend the stream network by as much as 40% an equation to measure the mean annual flood and its variability with stream density. The flows would almost double (1.96 times) where the road density was only 1.61 mi/mi2.

He notes that in many watersheds, particularly where attempts are being made to remove or obliterate older roads (including jammer-logging trails, major skid-trails), there are road densities that exceed 20 mi/mi2. Thus drainage efficiency can be expected to be an order of magnitude or more greater than on less roaded landscapes. Hollis found dramatic increases in the size of floods related to impervious surfaces.

He notes that small floods may be increased 10 times, and the 100-year flood may be doubled in size when drainage is basin-wide and 30% has been paved (left with impervious surfaces). When only 6% of the watershed was compacted significant peak flow increases, and emphasises that building and locating roads so as to not intercept and redirect water is very important. Shallow subsurface water can be converted to surface flow when intercepted by road cuts that exceed a few feet in height. Amounts vary by soil material, but surface flow volumes are increased substantially when sub-surface flows are intercepted by roads. It concluded that modifying road segments was the most effective way to approach watershed restoration.

Snowpack Issues Related to Use of Snowmobiles

Compacted snow melts rapidly and retains water less well than non-compacted snow. After compaction by snowmobiles, snow melting times increased to as much as double, and the potential water-holding capacity of snow compacted by snowmobiles was reduced substantially throughout the profile. During spring melt, these effects could significantly reduce the ability of snow to slow runoff and to moderate the effects of thawing.

Water Quality Issues Related to Backcountry Use

The relative expanse of most remote areas is misleading. Most use of remote areas is concentrated in the most accessible portions. Even in 1968, tons of chemicals were being introduced to the Boundary Waters Canoe Area, where 75% of the visitors entered the area through only 8 of the 66 entry points. National forest wilderness growth exceeds growth for many other forms of recreation taking place in the national forests. As a percentage of total national forest recreation use and of national forest campground use, wilderness use has grown steadily and only recently has shown a decline.

Continued population expansion and increased leisure time and affluence will require that wilderness areas absorb greater use.

WATER QUALITY ISSUES RELATED TO USE OF PERSONAL WATERCRAFT

Nearly all personal watercraft (PWC) utilise conventional two-stroke engines. As much as 30% of the fuel taken in by these engines is never used and is discharged, unburned, into water. As much as 4 out of 10 gallons of gasoline that are ingested by an outboard motor, may be discharged, unused, into the water. The use of leaded gasoline may result in lead accumulations in bottom mud that interrupt decomposition cycles and the food chain. Outboard motor gasoline also contains oil that contribute specific pollution problems to water: one gramme of oil requires 3.3 grammes of oxygen for complete oxidation, oil slicks interfere with gas exchange on the air-water surface, and oil also contains phosphates. The combustion process discharges additional toxic compounds into water. As a result, the use of PWCs has resulted in lower water quality in the nation's lakes and reservoirs (NPS 1999). Based on average use, a typical conventional two-stroke outboard or PWC will expel as much as 30% of the incoming fuel mixture, unburned, via the exhaust.

Also known as "two-cycle" engines, these motors intake a mixture of air, gasoline, and oil into the combustion chamber while exhaust gases are being expelled from the combustion chamber. Because the intake and exhaust processes are occurring at the same time, it is unavoidable that some of the unburned fuel mixture will escape with the exhaust. This expulsion of unburned fuel is the reason for the elevated levels of hydrocarbon emissions from conventional two-stroke engines. For example, at common fuel consumption rates, an average two-hour ride on a PWC may discharge three gallons of the gas-oil mixture into the water (NPS, 1999). Methyl tertiary butyl ether (MTBE) and polycyclic aromatic hydrocarbons (PAHs) are commonly observed twostroke engine contaminants and pose the most serious threats to human and ecological health. Water treatment facilities are generally ineffective in reducing MTBE concentrations.

Aquatic ecological communities do not appear to be threatened by observed concentrations of MTBE; however, more research is needed to reinforce this conclusion (NPS, 1999). The concentrations of PAH in lakes and reservoirs with high motorboat activity have been found at levels dangerous to aquatic organisms (NPS, 1999). The concentrations causing adverse effects can be extremely low (parts-pertrillion range) due to PAH phototoxicity, especially in oligotrophic waters where sunlight penetration is high.

The concentrations of PAH in lakes and reservoirs with high motorboat activity also have been found at levels dangerous to human health where humans are drinking the water and/or consuming the fish from these waters. Although PAH concentrations have not been widely measured, there is no reason to believe that the concentrations quoted are not widespread in lakes or reservoirs with high motorboat activity. Management strategies adopted by other agencies include outright bans on PWC and restricted use of two-stroke motors.

The exclusive use of the newly introduced and less polluting, direct-injection two-stroke engines has also been examined by water management agencies. One strategy for avoiding toxins in surface waters is to draw relatively uncontaminated water from deeper intervals to supply drinking water, but the consequences of this action upon amphibians, fish, insects, and algae has not been reported.

Chapter 7

Soil Pollution

Soil pollution comprises the pollution of soils with materials, mostly chemicals, that are out of place or are present at concentrations higher than normal which may have adverse effects on humans or other organisms. It is difficult to define soil pollution exactly because different opinions exist on how to characterise a pollutant; while some consider the use of pesticides acceptable if their effect does not exceed the intended result, others do not consider any use of pesticides or even chemical fertilizers acceptable. However, soil pollution is also caused by means other than the direct addition of xenobiotic (man-made) chemicals such as agricultural runoff waters, industrial waste materials, acidic precipitates, and radioactive fallout.

Both organic (those that contain carbon) and inorganic (those that don't) contaminants are important in soil. The most prominent chemical groups of organic contaminants are fuel hydrocarbons, polynuclear aromatic hydrocarbons (PAHs), polychlorinated biphenyls (PCBs), chlorinated aromatic compounds, detergents, and pesticides. Inorganic species include nitrates, phosphates, and heavy metals such as cadmium, chromium and lead; inorganic acids; and radionuclides (radioactive substances). Among the sources of these contaminants are agricultural runoffs, acidic precipitates, industrial waste materials, and radioactive fallout.

Soil pollution can lead to water pollution if toxic chemicals leach into groundwater, or if contaminated runoff reaches streams, lakes, or oceans. Soil also naturally contributes to air pollution by releasing volatile compounds into the atmosphere. Nitrogen escapes through ammonia volatilisation and denitrification. The decomposition of organic materials in soil can release sulfur dioxide and other sulfur compounds, causing acid rain. Heavy metals and other potentially toxic elements are the most serious soil pollutants in sewage. Sewage sludge contains heavy metals and, if applied repeatedly or in large amounts, the treated soil may accumulate heavy metals and consequently become unable to even support plant life.

In addition, chemicals that are not water soluble contaminate plants that grow on polluted soils, and they also tend to accumulate increasingly toward the top of the food chain. The banning of the pesticide DDT in the United States resulted from its tendency to become more and more concentrated as it moved from soil to worms or fish, and then to birds and their eggs. This occurred as creatures higher on the food chain ingested animals that were already contaminated with the pesticide from eating plants and other lower animals.

Lake Michigan, as an example, has 2 parts per trillion (ppt) of DDT in the water, 14 parts per billion (ppb) in the bottom mud, 410 ppb in amphipods (tiny water fleas and similar creatures), 3 to 6 parts per million (ppm) in fish such as coho salmon and lake trout, and as much as 99 ppm in herring gulls at the top of the food chain. The ever-increasing pollution of the environment has been one of the greatest concerns for science and the general public in the last fifty years.

The rapid industrialisation of agriculture, expansion of the chemical industry, and the need to generate cheap forms of energy has caused the continuous release of man-made organic chemicals into natural ecosystems. Consequently, the atmosphere, bodies of water, and many soil environments have become polluted by a large variety of toxic compounds. Many of these compounds at high concentrations or following prolonged exposure have the potential to produce adverse effects in humans and other organisms: These include the danger of acute toxicity, mutagenesis (genetic changes), carcinogenesis, and teratogenesis (birth defects) for humans and other organisms. Some of these man-made toxic compounds are also resistant to physical, chemical, or biological degradation and thus represent an environmental burden of considerable magnitude.

Numerous attempts are being made to decontaminate polluted soils, including an array of both *in situ* (on-site, in the soil) and off-site (removal of contaminated soil for treatment) techniques. None of these is ideal for remediating contaminated soils, and often, more than one of the techniques may be necessary to optimise the cleanup effort.

The most common decontamination method for polluted soils is to remove the soil and deposit it in landfills or to incinerate it. These methods, however, often exchange one problem for another: landfilling merely confines the polluted soil while doing little to decontaminate it, and incineration removes toxic organic chemicals from the soil, but subsequently releases them into the air, in the process causing air pollution.

For the removal and recovery of heavy metals various soil washing techniques have been developed including physical methods, such as attrition scrubbing and wet-screening, and chemical methods consisting of treatments with organic and inorganic acids, bases, salts and chelating agents. For example, chemicals used to extract radionuclides and toxic metals include hydrochloric, nitric, phosphoric and citric acids, sodium carbonate and sodium hydroxide and the chelating agents EDTA and DTPA. The problem with these methods, however, is again that they generate secondary waste products that may require additional hazardous waste treatments.

In contrast to the previously described methods, *in situ* methods are used directly at the contamination site. In this case, soil does not need to be excavated, and therefore the chance of causing further environmental harm is minimised. *In situ* biodegradation involves the enhancement of naturally occurring microorganisms by artificially stimulating their numbers and activity. The microorganisms then assist in degrading the soil contaminants.

A number of environmental, chemical, and management factors affect the biodegradation of soil pollutants, including moisture content, pH, temperature, the microbial community that is present, and the availability of nutrients. Biodegradation is facilitated by aerobic soil conditions and soil pH in the neutral range (between pH 5.5 to 8.0), with an optimum reading occurring at approximately pH 7, and a temperature in the range of 20 to 30°C. These physical parameters can be influenced, thereby promoting the microorganisms' ability to degrade chemical contaminants. Of all the decontamination methods bioremediation appears to be the least damaging and most environmentally acceptable technique.

ABATEMENT

Abatement is a general term used for methods or technologies that reduce the amount of pollutant generated in a chemical or other manufacturing facility. In contrast, the terms cleanup and remediation refer to removal or appropriate disposal of the pollutants after they have been generated; these methods are also often referred to as *end-of-the-pipe treatment*. Current industrial practice places more emphasis on abatement (also known as pollution prevention) and follows the following simple rule: "If you don't make it, you don't treat it."

Pollution abatement involves source reduction, in-process recycling, in-plant recycling, design modifications, off-site recycling, and treatment to make the waste less hazardous. Source reduction refers to the examination of various processing units in detail to determine if wastes can be minimised. The step involves several layers of study:

- Waste inventory is generated.
- Critical processes leading to waste are identified.
- Alternative processing strategies are studied to reduce the amount of waste generated in these processes. The collection of waste inventory is an important part of such an analysis. In

addition, the inputs that generate these wastes are identified. These data then suggest ideas for source reduction.

In batch reactors, for example, especially in the manufacture of dyestuffs, rinsing of the reactors in between batches is needed to avoid the contamination of the product made in the next batch. This generates a stream of wastewater. The quantity can be reduced by optimal batch scheduling, that is, making similar dyes for a while before switching to a different colour. This requires less rinsing between batches since the next dye to be made is of the same colour. Another example is the use of solvents. Contaminated solvent from one part of the process may still be good enough for another part of the plant, and the overall generation of the contaminated solvent can be reduced by properly identifying the solvent needs of the entire plant.

In-process recycling refers to the reuse of unreacted materials (after suitable purification) in the same process. Chemical reactions, for example, cannot always be driven to completion due to *thermodynamic limitations*. In such cases, one needs to separate the product from the unconverted raw material, the latter ending up as a waste stream. The waste generation is abated by recycling the raw materials back to the process. The recycling of spent solvent after some needed purification (e.g., carbon adsorption or steam stripping) is another example of this strategy.

In-plant recycling refers to the use of waste generated in one part of the production as a raw material for another part of the plant. Examples of in-plant recycling include solvent reuse and water reuse in the chemical industry.

Design modifications play an important role in waste minimisation. Often, minor modifications in the existing equipment can result in considerable waste reduction. The better design of cyclone separators can reduce the number of dust particles exiting a process. For storage tanks, floating roofs are often used in place of fixed roof tanks to avoid "breathing" losses, or losses from a tank when the ambient conditions (temperature and pressure) change. For example, if the atmospheric pressure drops, some vapour escapes from the tank in order to equilibrate the pressure, leading to air pollution. Proper insulation can reduce the waste sludge formation in distillation column reboilers because these may now be operated at a lower temperature. A distillation column reboiler is a component of a distillation column and is widely used in refineries and other chemical plants.

Off-site recycling applies to the situation where the waste generated in one plant is a raw material for another industry. For example, gypsum (calcium sulfate) is a waste from stack gas (sulfur dioxide) cleaner in the coal industry, but a raw material in the cement industry. The proximity of industries is an important consideration in off-site recycling since

often involve the use of reactive agents such as sulfur to create immobile sulfide salts of metals (e.g., mercury). Reducing the mobility of soil contaminants often refers to reducing the water solubility of the compounds. Reducing water solubility lowers the potential for contaminants to become dissolved in and move with water in the subsurface.

Groundwater

DNAPLs such as chlorinated solvents may be treated with chemicals (e.g., potassium permanganate) that degrade the solvents into relatively harmless chemicals. When combined with chlorinated solvents, potassium permanganate removes chloride ions, which results in the degradation of these chemicals to carbon dioxide (CO_2) and water. This technology holds promise as a tool for remediating these challenging contaminants.

CONTAINMENT

Situations exist in which technologies are not available or practical to remove or convert contaminants. In those situations, it is often possible to contain the contamination as a final solution or as an interim measure until appropriate technologies become available. Soils. Radionuclides from historical weapons production and nuclear testing, as well as from industrial uses of radiation, appear to be a good match for developing containment technologies.

For example, containment is a promising technology for the management of radioactively contaminated soils beneath the large high-level radioactive waste storage tanks at the US Department of Energy Hanford site in Washington State. Removing radioactive contamination from soil is problematic from a worker-safety standpoint, and it may create further contamination of equipment, containers, and surrounding areas. Efforts to develop effective physical containment technologies for soil contaminants are continuous.

Groundwater

Groundwater is not generally suitable for absolute containment; however, between containment and conversion is a technology known as reactive barriers. Reactive barriers intercept contaminated groundwater plumes and are constructed of chemically reactive materials (e.g., iron) that bind or convert dissolved contaminants. Reactions between the contaminant and the iron either immobilise or degrade the contaminant by altering its chemical form (redox manipulation).

LAND POLLUTION

Land pollution is the degradation of earth's land surfaces often caused by human activities and its misuse. Haphazard disposal of urban and industrial wastes, exploitation of minerals, and improper use of soil by inadequate agricultural practices are a few of the contributing factors. Also, increasing urbanisation, industrialisation and other demands on the environment and its resources is of great consequence to many countries. The Industrial Revolution set in motion a series of events which impinged on the countryside destroying many natural habitats, and introduced pollution causing disease in both human and animal alike.

Increased Mechanisation

In some areas more metal ores had to be extracted out of the ground, melted and cast using coal out of the ground and cooled using water, which raised the temperature of water in rivers. (This reduces the oxygen carrying capacity of the water and affects all the living things there.) The excavation of metal ores, sand and limestone led to large scale quarrying and defacing of the countryside.

To a large extent this has stopped or is more closely controlled, and attempts have been made to use the holes profitably i.e., sand pits have been turned into boating lochs and quarries have been used as landfill waste sites. Central Scotland bears the scars of years of coal mining, with pit binges and slag heaps visible from the motorways.

Increased Urbanisation

As the demand for labour grew, the areas round the factories and mines were given over to housing. This took up former agricultural land, caused sewage and waste problems, increased the demands for food and put pressure on farmers to produce more food.

The demand for more housing meant the need to use more raw materials to make bricks, slates for roofing and timber for joists, etc. Once again this led to quarrying and to the destruction of forests. The houses also needed running water and a supply of energy. Initially this water would have been supplied directly from a stream but as demand increased the need for reservoirs increased. This again led to the loss of land as valleys were flooded to meet the demands. The main fuels used would have been coal and wood but as time progressed, hydro electric, coal, oil and nuclear power stations were built which again became features or eyesores on the landscape.

Associated with this was the radiating network of pylons forming the National Grid, as well as, the substations and transformers. Until the late 1970s little attempt was made to hide these metal structures but now more care is taken in their sitting and underground cables are often used- although these are not popular with repair crews who have to find faults and service them, often in very remote areas. This increase in the concentration of population into cities, along with the internal combustion engine, led to the increased number of roads and all the infra structure that goes with them. Roads cause visual, noise, light, air and water pollution, as well as using up land. The visual and noise areas are obvious, however light pollution is becoming more widely recognised as a problem.

From space large cities can be picked out at night by the glow of their street lighting, so city dwellers seldom experience total darkness. On a smaller scale lights along roads can cause people living there to have interrupted sleep patterns due to the lack of darkness.

The contribution of traffic to air pollution is dealt with in another article, but, suffice to say that sulfur dioxide, nitrogen oxides and carbon monoxide are the main culprits. Water pollution is caused by the run off from roads of oil, salt and rubber residue, which enter the water courses and may make conditions unsuitable for certain organisms to live.

Increased Agricultural Land and Field Size

As the demand for food has grown so high, there is an increase in field size and mechanisation. The increase in field size is to make it economically viable for the farmer but results in loss of habitat and shelter for wildlife as hedgerows and copses disappear. When crops are harvested the naked soil is left open to wind blow after the heavy machinery has crossed and compacted it. Another consequence of more intensive agriculture is the move to monoculture. This is unnatural, it depletes the soil of nutrients, allows diseases and pests to spread and, in short, brings into play the use of chemical substances foreign to the environment.

Pesticides

Pesticides are any chemical used to remove pests whether they are plants or animals. They are used to kill wire worms and slugs that attack cereal crops and to kill ergot – *Claviceps purpurea* – a fungus that attacks crops and may get into human food.

Herbicides

Herbicides are used to kill weeds, especially on pavements and railways. They are similar to auxins and most are biodegradable by soil bacteria. However, one group derived from trinitrophenol (2:4 D and 2:4:5 T) have the impurity dioxin which is very toxic and causes fatality even in low concentrations.

It also causes spontaneous abortions, haemorrhaging and cancer. Agent Orange (50% 2:4:5 T) was used as a defoliant in Vietnam. Eleven million gallons were used and children born since then to American soldiers who served in this conflict, have shown increased physical and mental disabilities compared to the rest of the population. It affects the head of the sperm and the chromosomes inside it. Another herbicide, much loved by murder story writers, is Paraquat. It is highly toxic but it rapidly degrades in soil due to the action of bacteria and does not kill soil fauna.

Fungicides

Fungicides are the group used to stop the growth of smuts and rusts on cereals, and mildews and moulds like Mucor on plants. The problem is that they may contain copper and mercury. Copper is very toxic at 1ppm to water plants and fish and can enter the skin if being sprayed to reduce mildew and accumulate in the central nervous system. Organomercury compounds have been used to get rid of sedges which are insidious and difficult to remove. However, it also can accumulate in birds' central nervous system and kill them.

Insecticides

Insecticides are used to rid farmers of pests which damage crops. The insects damage not only standing crops but also stored ones and in the tropics it is reckoned that one third of the total production is lost during food storage. As with fungicides, the first used in the nineteenth century were inorganic e.g., Paris Green and other compounds of arsenic. Nicotine has also been used since the late eighteenth century. There are now two main groups of synthetic ones:

Organochlorines

Organochlorines include DDT, Aldrin, Dieldrin and BHC. They are cheap to produce, potent and persistent. DDT was used on a massive scale from the 1930s, with a peak of 72,000 tonnes used 1970. Then usage fell as the environmental problems were realised. It was found worldwide in fish and birds and was even discovered in the snow in the Antarctic. It is only slightly soluble in water but is very soluble in the bloodstream. It affects the nervous and enzyme systems and causes the eggshells of birds to lack calcium and be so fragile that they break easily. It is thought to be responsible for the decline of the numbers of birds of prey like ospreys and peregrine falcons in the 1950s – they are now recovering.

As well as increased concentration via the food chain, it is known to enter via permeable membranes, so fish get it through their gills. As it has low solubility it tends to stay at the surface, so organisms that live there are most affected. DDT found in fish that formed part of the human food chain caused concern but the levels found in the liver, kidney and brain tissues was less than 1ppm and in fat was 10 ppm which was below the level likely to cause harm. However, DDT was banned in Britain and

America to stop the further building up of it in the food chain. However, the USA exploited this ban and sold DDT to developing countries who could not afford the expensive replacement chemicals and who did not have such stringent regulations governing the use of pesticides.

Some insects have developed a resistance to insecticides – e.g., the Anopheles mosquito which carries malaria.

Organophosphates

Organophosphates, e.g. parathion, methyl parathion and about 40 other insecticides are available nationally. Parathion is highly toxic, methyl-parathion is less so and Malathion is generally considered safe as it has low toxicity and is rapidly broken down in the mammalian liver. This group works by preventing normal nerve transmission as cholinesterase is prevented from breaking down the transmitter substance acetylcholine, resulting in uncontrolled muscle movements.

Entry of a variety of pesticides into our water supplies causes concern to environmental groups, as in many cases the long term effects of these specific chemicals is not known.

Limits came into force in July 1985 and were so frequently broken that in 1987 formal proceedings were taken against the British Government. Britain is still the only European state to use Aldrin and organ chlorines, although it was supposed to stop in 1993. East Anglia has the worst record for pesticide contamination of drinking water. Of the 350 pesticides used in Britain, only 50 can be analysed – this is a worrying thought for many people.

Increased Waste Disposal

In Scotland in 1993, 14 million tons of waste were produced. 100,000 tons were special waste and 260,000 tons were controlled waste from other parts of Britain and abroad. 45% of the special waste were in liquid form and 18% were asbestos - radioactive waste was not included. Of the controlled waste, 48% comes from the demolition of buildings, 22% from industry, 17% from households and 13% from business – only 3% are recycled. 90% of controlled waste are buried in landfill sites and produces 2 million tons of methane gas. 1.5% is burned in incinerators and 1.5% are exported to be disposed of or recycled. There are 748 disposal sites in Scotland.

Landfill produces leachate, which has to be recycled to keep favourable conditions for microbial activity, and methane gas and some carbon dioxide. There are very little contaminated vacant or derelict land in the north east of Scotland as there are little traditional heavy industry or coal/mineral extraction. However, some soil are contaminated by aromatic hydrocarbons (500 cubic meters). The Urban Waste Water Treatment Directive allows sewage sludge to be sprayed onto land and the volume is expected to double to 185,000 tons of dry solids in 2005. This has good agricultural properties due to the high nitrogen and phosphate content. In 1990/1991, 13% wet weight was sprayed onto 0.13% of the land, however this is expected to rise 15 folds by 2005. There is a need to control this so that pathogenic microorganisms do not get into water courses and to ensure that there are no accumulation of heavy metals in the top soil.

Increased Leisure and Available Wealth

At the end of twentieth century people had even more leisure time and available wealth. This means that people can travelled around the countryside more often increasing the number of cars. This is related back to the roads issue but has also led to the increased litter problem in the countryside. This is usually packaging, cans, bottles, etc., from picnics but increasingly people are dumping household rubbish in the countryside instead of taking it to the local tip. Aesthetically litter is unpleasant but poses threats to the wildlife through razor sharp glass that can be trodden on, plastic bags that can be eaten, etc. More and more litter is becoming a problem especially in the more remote areas which are now more accessible to the general public. Until the public take responsibility to stop littering, then legislation will have little effect and information and education will be the fore runners in the fight against the litter bugs.

Increased Military Presence

As nations grow so do their armed forces. Over the century, the army, the navy and latterly the air force has grown in Britain and so has their ownership of land. Apart from the noise and aviation fuel pollution of the air bases, the destruction of land on firing ranges and the change in coastlines to form naval bases, a more sinister trend is the increase in research stations with their "hidden agendas and experiments".

Anthrax is caused by the bacterium *Bacillus anthracis*. It was discovered in the 1870s by the German scientist Robert Koch. It mainly affects herbivores, causing them to stagger, convulse and die in a few days. It can also affect man if the spores get onto the skin or lungs.

It will form a pus filled blister and was initially treated by a vaccine prepared by Louis Pasteur in 1881. When an animal has died of the disease, the only safe way to dispose of it is to burn it or to bury it very deep in the earth. However in World War II, knowing all the above problems, the British Government decided to use Anthrax as a biological weapon. In 1942 they dropped Anthrax bombs on Gruinard Island. Their idea – and indeed they produced these – was to drop 5 million Anthrax inoculated linseed cakes into fields of German cattle. The cattle cakes were destroyed at the end of the war unused.

However, the Anthrax spores on Gruinard persisted for 40 years until in 1986 the whole island was decontaminated by formaldehyde, and in 1990 returned to its original owners.This was an example of short sightedness that cost the island of Gruinard 50 years of its "natural life" and which could have spread out of control had it been used on mainland Europe.

SOIL CONTAMINATION

Soil contamination is caused by the presence of man-made chemicals or other alteration in the natural soil environment. This type of contamination typically arises from the rupture of underground storage tanks, application of pesticides, percolation of contaminated surface water to subsurface strata, oil and fuel dumping, leaching of wastes from landfills or direct discharge of industrial wastes to the soil. The most common chemicals involved are petroleum hydrocarbons, solvents, pesticides, lead and other heavy metals.

This occurrence of this phenomenon is correlated with the degree of industrialisation and intensity of chemical usage.The concern over soil contamination stems primarily from health risks, both of direct contact and from secondary contamination of water supplies. Mapping of contaminated soil sites and the resulting cleanup are time consuming and expensive tasks, requiring extensive amounts of geology, hydrology, chemistry and computer modeling skills.

It is in North America and Western Europe that the extent of contaminated land is most well known, with many of countries in these areas having a legal framework to identify and deal with this environmental problem; this however may well be just the tip of the iceberg with developing countries very likely to be the next generation of new soil contamination cases.

The immense and sustained growth of the People's Republic of China since the 1970s has exacted a price from the land in increased soil pollution. The State Environmental Protection Administration believes it to be a threat to the environment, to food safety and to sustainable agriculture. According to a scientific sampling, 150 million mi (100,000 square kilometres) of China's cultivated land have been polluted, with contaminated water being used to irrigate a further 32.5 million mi (21,670 square kilometres) and another 2 million mi (1,300 square kilometres) covered or destroyed by solid waste. In total, the area accounts for one-tenth of China's cultivatable land, and is mostly in economically developed areas.

The United States, while having some of the most widespread soil contamination, has actually been a leader in defining and implementing standards for cleanup. Other industrialised countries have a large number of contaminated sites, but lag the US in executing remediation. Developing countries may be leading in the next generation of new soil contamination cases.

Each year in the US, thousands of sites complete soil contamination cleanup, some by using microbes that "eat up" toxic chemicals in soil, many others by simple excavation and others by more expensive high-tech soil vapour extraction or air stripping.

At the same time, efforts proceed worldwide in creating and identifying new sites of soil contamination, particularly in industrial countries other than the US, and in developing countries which lack the money and the technology to adequately protect soil resources.

Microanalysis of Soil Contamination

To understand the fundamental nature of soil contamination, it is necessary to envision the variety of mechanisms for pollutants to become entrained in soil.

Soil particulates may be composed of a gamut of organic and inorganic chemicals with variations in cation exchange capacity, buffering capacity, and redox poise. For example, at the extremes, one has a sand component, a coarse grained, inert, and totally inorganic substance; whereas peat soils are dominated by a fine organic material, made of decomposing organic material and highly active.

Most soils are mixtures of soil subtypes and thus have quite complex characteristics. There is also a great diversity of soil porosity, ranging from gravels to sands to silt to clay (in increasing order of porosity), pore size, and pore tortuosity (both in decreasing order). Finally, there is a wide spectrum of chemical bonding or adhesion characteristics: each contaminant has a different interaction or bonding mechanism with a given soil type.

On balance, some contaminants may literally drain through soils such as sand and gravel and move to other soils or deeper aquifers, while polar or organic chemicals discharged into a clay soil will have a very high adsorption. Thus most soil contamination is the result of pollutants adhering to the soil particle surface, or lodging in interstices of a soil matrix.

Clearly the equilibrium reached is a dynamic one, where new pollutants may lodge on new soil particles and the action of groundwater movement may over time transport some of the soil contaminants to other locations or depths. Soil contamination results when hazardous substances are either spilled or buried directly in soil or migrate to soil from a spill that has occurred elsewhere. For example, soil can become contaminated when small particles containing hazardous substances are released from a smokestack and are deposited on the surrounding soil as they fall out of the air.

Another source of soil contamination could be water that washes contamination from an area containing hazardous substances and deposits the contamination in the soil as it flows over or through it.

Health Efects

The major concern is that there are many sensitive land uses where people are in direct contact with soils such as residences, parks, schools and playgrounds. Other contact mechanisms include contamination of drinking water or inhalation of soil contaminants which have vaporised. There is a very large set of health consequences from exposure to soil contamination depending on pollutant type, pathway of attack and vulnerability of the exposed population. Chromium and obsolete pesticide formulations are carcinogenic to populations.

Lead is especially hazardous to young children, in which group there is a high risk of developmental damage to the brain,while to all populations kidney damage is a risk.Chronic exposure to at sufficient concentrations is known to be associated with higher incidence of leukemia.

Obsolete pesticides such as mercury and cyclodienes are known to induce higher incidences of kidney damage, some irreversible; cyclodienes are linked to liver toxicity.

Organophosphates and carbamates can induce a chain of responses leading to neuromuscular blockage. Many chlorinated solvents induce liver changes, kidney changes and depression of the central nervous system. There is an entire spectrum of further health effects such as headache, nausea, fatigue (physical), eye irritation and skin rash for the above cited and other chemicals.

Ecosystem Efects

Not unexpectedly, soil contaminants can have significant deleterious consequences for ecosystems. There are radical soil chemistry changes which can arise from the presence of many hazardous chemicals even at low concentration of the contaminant species.

These changes can manifest in the alteration of metabolism of endemic microorganisms and arthropods resident in a given soil environment. The result can be virtual eradication of some of the primary food chain, which in turn have major consequences for predator or consumer species.

Even if the chemical effect on lower life forms is small, the lower pyramid levels of the food chain may ingest alien chemicals, which normally become more concentrated for each consuming rung of the food chain. Many of these effects are now well known, such as the concentration of persistent DDT materials for avian consumers, leading to weakening of egg shells, increased chick mortality and potentially species extinction.Effects occur to agricultural lands which have certain types of soil contamination.

Contaminants typically alter plant metabolism, most commonly to reduce crop yields. This has a secondary effect upon soil conservation, since the languishing crops cannot shield the earth's soil mantle from erosion phenomena. Some of these chemical contaminants have long half-lives and in other cases derivative chemicals are formed from decay of primary soil contaminants.

Regulatory Framework

Until about 1970 there was little widespread awareness of the worldwide scope of soil contamination or its health risks. In fact, areas of concern were often viewed as unusual or isolated incidents. Since then, the US has established guidelines for handling hazardous waste and the cleanup of soil pollution.

In 1980 the US Superfund/CERCLA established strict rules on legal liability for soil contamination. Not only did CERCLA stimulate identification and cleanup of thousands of sites, but it raised awareness of property buyers and sellers to make soil pollution a focal issue of land use and management practices.While estimates of remaining soil cleanup in the US may exceed 200,000 sites, in other industrialised countries there is a lag of identification and cleanup functions.

Even though their use of chemicals is lower than industrialized countries, often their controls and regulatory framework is quite weak. For example, some persistent pesticides that have been banned in the US are in widespread uncontrolled use in developing countries. It is worth noting that the cost of cleaning up a soil contaminated site can range from as little as about $10,000 for a small spill, which can be simply excavated, to millions of dollars for a widespread event, especially for a chemical that is very mobile such as perchloroethylene.

Cleanup Options

Cleanup or remediation is analysed by environmental scientists who utilise field measurement of soil chemicals and also apply computer models for analysing transport and fate of soil chemicals. Thousands of soil contamination cases are currently in active cleanup across the US as of 2006. There are several principal strategies for remediation:

- Excavate soil and remove it to a disposal site away from ready pathways for human or sensitive ecosystem contact. This technique also applies to dredging of bay muds containing toxins.

- Aeration of soils at the contaminated site (with attendant risk of creating air pollution).
- Bioremediation, involving microbial digestion of certain organic chemicals. Techniques used in bioremediation include landfarming, biostimulation and bioaugmentation soil biota with commercially available microflora.
- Extraction of groundwater or soil vapour with an active electromechanical system, with subsequent stripping of the contaminants from the extract.
- Containment of the soil contaminants.

Chapter 8

Industrial Waste

REDUCTION OF FLUORINE-CONTAINING INDUSTRIAL WASTE USING ALUMINUM-SOLUBILITY METHOD

Recently environmental protection is being given a great deal of attention, and standards to control the concentration of substances discharged from factories is becoming more strict every year. The concentration of fluorine in waste water, for example, is defined as 15 mg/l or less by the Water Pollution Control Law. This value is even 8 mg/l or less depending on the local self governing body. Each factory performs the treatment of waste water and exhaust gas by various methods to meet these control values. These treatments generate an enormous amount of sludge, which are reclaimed as industrial waste. But the capacity of reclamation plants is already saturated, and decreasing industrial waste is an urgent problem. Semiconductor plants are no exception, where large amounts of waste sludge is discharged. About 30% of waste sludge is generated when treating fluorine-containing waste water.

To decrease the generation of this waste, various treatments have been examined. In newly established waste water treatment plants, a new system which uses calcium carbonate, is having good results. We examined the treatment method using an aluminumsolubility method to decrease the fluorine-containing waste water treatment sludge generated from semiconductor plants. This chapter first describes the current situation of fluorine-containing waste water treatment, then the principles, experimental methods and performance of treatment methods using an aluminum-solubility method are described.

CURRENT SITUATION OF FLUORINE-CONTAINING WASTE WATER TREATMENT

Coagulating Sedimentation Method Using Calcium Hydroxide

The primary fluorine-containing waste water treatment method conventionally used was the coagulating sediment method which uses calcium hydroxide {$Ca(OH)_2$} as the treatment agent.

- *First process* (first reaction tank): Calcium hydroxide is added to fluorine-containing waste water, and the pH is controlled. Fluorine ions that dissolve in the waste water combine with calcium ions and is deposited as calcium fluoride.
- *Second process* (first coagulating sedimentation tank): Since calcium fluoride is alkaline and exhibits low solubility, solid and liquid can be easily separated by using the coagulating sedimentation method. The separation speed can be accelerated by adding macromolecular flocculant. Supernatant liquid after solid-liquid separation is discharged as treated water and the sedimentated portion is recovered as sludge.
- *Third process* (dehydrator): Recovered sludge is decreased in volume by a dehydrator and is disposed of as industrial waste.

Problems of Conventional Methods

Calcium hydroxide, which is frequently sold as slaked lime, is generally inexpensive. But the solubility of calcium hydroxide is low, so the quantity of calcium hydroxide that dissolves in water as calcium ions is very little.

Therefore, a large amount of the treatment agent must be added to treat the concentration of the treatment target fluorine. Excess calcium hydroxide added in the waste water is recovered with calcium fluoride as waste sludge, which is a cause to increase waste sludge more than necessary.

Also because of low solubility, calcium hydroxide can decrease the concentration of fluorine in treated water only to 15 mg/l. To further decrease this concentration to 8 mg/l or less, a secondary treatment (second reaction tank - second coagulating sedimentation tank) using an aluminum treatment agent is required.

REDUCTION OF FLUORINE-CONTAINING WASTE SLUDGE

Concept

Under certain conditions, aluminum exhibits a higher bonding strength with fluorine compared with calcium. Aluminum can therefore remove fluorine in waste water down to a lower concentration. The problems of aluminum are that aluminum is much more expensive than the calcium treatment agent, and sludge generated after treatment has a poor dehydrating capability.

For these reasons, aluminum is inappropriate for high concentration fluorine treatment, and therefore is mainly used for low concentration fluorine waste water. Focusing on the high fluorine removal capability of aluminum, and the good dehydrating capability and low cost of calcium, we examined a treatment method that decreases waste sludge.

WASTE WATER TREATMENT USING ALUMINUM-SOLUBILITY METHOD

We experimented with a waste water treatment based on an aluminum-solubility method, and checked the effect. With this method, a calcium treatment agent is added to the sludge generated after treating fluorine-containing waste water with an aluminum treatment agent, and the aluminum of aluminum fluoride is substituted with calcium. Aluminum in sludge dissolves, and is recovered and reused as an aluminum treatment agent. This waste water treatment system is separated into five processes.

First Process

The recycled aluminum treatment agent, recovered in the fourth process, is added to the fluorine-containing waste water. When this is insufficient for the fluorine treatment, a new aluminum treatment agent is added.

By pH control, fluorine contained in waste water is deposited as aluminum fluoride, then fluorine in waste water is removed by removing aluminum fluoride. The following chemical formula shows the reaction of hydrofluoric acid and ammonium fluoride with the aluminum treatment agent.

$$3\,HF + AlO_2 \xrightarrow{-NaOH} AlF_3\downarrow + H_2O + OH^-$$

$$3\,NH_4F + AlO_2 \xrightarrow{-NaOH} AlF_3\downarrow + 3NH_4^+ + O_2$$

$$6\,NH_4HF_2 + 4\,AlO_2^- \xrightarrow{-NaOH} 4\,AlF_3\downarrow + 6\,NH_4^+ + 2\,O_2 + 2\,H_2O + 2\,OH^-$$

The relationship between the volume of deposit of aluminum fluoride and the concentration of fluorine in treated water obtained from our experiment.

Second Treatment

Recovered sludge is separated from treated water by solidliquid separation.

We measured the concentration of aluminum in recovered sludge, and confirmed that 85% of aluminum can be recovered.

Third Process

A calcium treatment agent is added to the recovered sludge. By pH control, aluminum of aluminum fluoride is substituted with calcium and is deposited as calcium fluoride. The following formula shows the reaction between aluminum fluoride sludge and the calciumtreatment agent.

$$2\,AlF_3\downarrow + 3Ca^{2+} \xrightarrow{-NaOH} 3CaF_4\downarrow + 2\,AlO_2^- + 4H^+ + 2Na^+$$

When aluminum fluoride dissolves and fluorine becomes unstable, aluminum of aluminum fluoride is substituted with calcium and is deposited as calcium fluoride.

The theoretical quantity of adding calcium is calcium: fluorine = 1:2. Based on the above result, we determined the quantity of the aluminum treatment agent and the calcium treatment agent to add that is required for the removal of fluoride and for substitution.

RADIOACTIVE WASTE: THE PROBLEM AND ITS MANAGEMENT

Radioactive Waste

Two basic nuclear reactions, namely fission of nuclei like 235U, 239Pu and fusion of elements like hydrogen result in release of enormous energy and radioactive elements. Controlled vast releases of energy are possible in nuclear power plant reactors through the fission reaction. The dream of controlled vast releases of energy through fusion reaction is still to be realised.

Uncontrolled vast releases of energy through both these reactions have been possible in "atom" and "hydrogen" (thermonuclear) bombs. As in many other industrial processes, in the nuclear industry also, one gets unusable and unwanted *waste* products; the residues turn out to be hazardous.

Waste, by definition, is any material (solid materials such as process residues as well as liquid and gaseous effluents) that has been or will be discarded *as being of no further use*. Note that what may be considered as one's waste may turn out to be another's wealth. Reusable plastics and other components in day-to-day household waste are good examples in this context. This concept holds good for radioactive waste also, in some sense. Waste that emits nuclear radiation is radioactive waste.

Natural Radioactivity

It is somewhat surprising that nature has been a large producer of radioactive waste. Over the eons, the surface of the Earth and the terrestrial crust happens to be an enormous reservoir of primordial radioactivity. Small amounts of radioactive materials are contained in mineral springs, sand mounds and volcanic eruptions.

Essentially, all substances contain radioactive elements of natural origin to some extent or the other. The second source of radioactive waste is a part of industrial mining activity where, during mineral exploration and exploitation, one excavates the primordial material from the Earth that contains radioactivity, uses part of it and rejects the radioactive residues as waste.

These are referred to as Naturally Occurring Radioactive Materials (NORMs) and are ubiquitous as residual wastes in processing industries that cover fertilizers, iron and steel, fossil fuel, cement, mineral sands, titanium, thorium and uranium mining as well as emanations and waste from coal and gas-fired power plants.

One should note that in many industries, radiation exposure to the workers and the general public would be at least as high as those from nuclear installations and in some cases it is even higher. It is also known that certain mineral springs contain fairly large amounts of radon. Monazite sand deposits in coastal areas may result in radiation exposure to humans around an order of magnitude in excess of the currently set international exposure limits to radioactive waste disposal (one msv/year) and volcanic deposits result in similar exposure.

There is no place on Earth that is free from natural radioactive background; it may vary from place to place all the way from the low to the high. The content of radioactivity in the seas is estimated to be nearly 10,000 exabecquerel (Ebq = 1018 Bq). The residual waste tailings from past mining and milling operations are estimated to be around several million tons at many places and the radioactivity contained may be nearly 0.001 EBq.

Thousands of such sites are scattered all around the world. At OKLO, located in Gabon in the West African rainforest, there exists uranium ore that formed an active natural reactor over some billion years ago.

Artificial Radioactivity

Radioactivity was discovered about a hundred years ago. Following the Second World War and discovery of the fission process, human activity added radioactivity artificially to the natural one. Two main sources have been: (1) the civilian nuclear programmes, including nuclear power production, medical and industrial applications of radioactive nuclides for peaceful purposes, and (2) the military nuclear programme, including atmospheric and underground nuclear-weapon testing and weapon production.

Nuclear Fuel Cycle

As stated earlier, civilian nuclear operations lead to radioactivity. The story of uranium from its mining to its use in reactors and thence of chemical processing and accumulation of radioactive waste is covered by what is referred to as "nuclear fuel cycle". The ore that is mined in uranium mines is sent to a uranium mill, where a small uranium-containing fraction is separated from the ore, leaving behind virtually almost the entire ore in the tailings.

The uranium fraction is processed to recover pure uranium in metallic form. Uranium metal consists of the isotope 235U to the extent of 0.7%, the remaining 99.3% being 238U. 235U fissions on absorption of thermal neutrons, while 238U does not.

Hence this small fraction of 235U is "enriched" for use in light-water reactors for deriving power. Highly enriched 235U is used for nuclear weapons also. In CANDU-type heavy-water reactors, one can use natural uranium itself as fuel, without any enrichment.

Fresh fuel made of uranium (sometimes containing plutonium, in addition) is weakly radioactive. The fuel, after sufficient use in reactors, is referred as "spent fuel"; the "ash" after "burning" the fuel contains fissionfragment debris from spontaneous or neutron-induced fission of uranium and actinides, actinide elements and unutilised uranium.

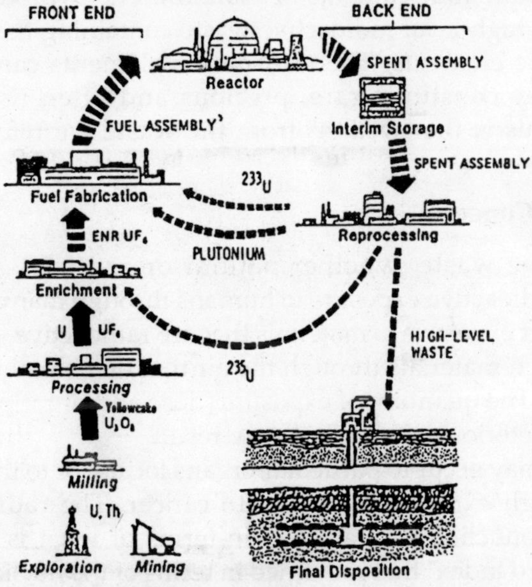

Fig:1. Nuclear Fuel Cycle

This irradiated fuel is highly radioactive. Transuranic actinides (principally neptunium, plutonium, americium and curium) are created by absorption of neutrons in non-fissioned uranium and by sequential absorption of neutrons in the consequently formed daughter elements. Although nearly 200 radionuclides are produced during the burn-up of the fuel, the great majority of them are relatively short-lived and decay to low levels within a few decades. Hence the spent fuel is often allowed to "cool" in spentfuel bays of water, to allow short-lived radioactivity to decay. Often such fuel is stored for indefinite time in the fuel pools without any further processing, or in dry "coffins".

The short-lived radionuclides therefore do not pose a big problem for long-term disposal. The spent fuel, when subjected to chemical processing, yields uranium and plutonium fractions apart from the rest of the "ash". In this article we do not deal with the chemistry involving a variety of highly toxic chemicals or the complex chemical processes that one employs, either in fuel reprocessing or in radioactive waste management.

Beginning with dissolution of cladded burnt fuel to retrieving useful fissile elements is an enormous activity involving chemical engineering, remote handling, monitoring, etc. As opposed to the socalled "once-through fuel cycle" wherein no material is recycled, in the "closed-cycle fuel cycle", uranium is recycled for fuel production and plutonium for either fuel production or for weapons. Normally, the irradiated uranium is dissolved in an acid medium and treated with organic solvents to recover plutonium and remnants of uranium. The byproduct is a highly acidic liquid, a high-level radioactive waste containing fission fragments and transuranic elements. The transuranic elements can be separated further, as they constitute rare, precious and often fissile materials themselves. This is the "wealth" from the waste we referred to in the beginning.

The Cause for Concern

Radioactive waste, whether natural or artificial, is a potential harbinger of radioactive exposure to humans through many channels. The routes are direct exposure to materials that are radioactive, inhalation and ingestion of such materials through the air that one breathes or food that one consumes. The quantum of exposure (dose and duration of exposure) decides the deleterious effects that may result.

Exposure may occur to particular organs locally or to the whole body. Sufficiently high exposure can lead to cancer. The radiotoxicity of a particular radionuclide is quantified in terms of what is referred to as "potential hazard index" that is defined in terms of the nuclide availability, its activity, maximum permissible intake annually and its half-life.

This depends on a variety of factors like physical half-life, biological half-life, sensitivity of the organ or tissue where the nuclide is likely to concentrate, ionizing power of the radiation from the nuclide that depends on the energy of the radiation emitted from the radionuclide, etc. It is from such considerations that one concludes that radioactive nuclides of elements like 137Cs or 90Sr or 131I are the most hazardous on the scale of a human beings' lifetime. Other long-life nuclides like 239Pu, 241Am, 237Np pose a long-term hazard, on the other hand, to future generations.

Although nature's sources are to be as much feared as those from artificial sources, the (atomic) bomb's legacy has set a certain perception in the public mind, of the dangers inherent or implicit in the use and abuse of nuclear facilities, operations and waste.

The recent emphasis arises because of concern to the effects on the environment over a very long period of time. High-level radioactive waste is potentially toxic for tens of thousands to millions of years; it is also the most difficult to be disposed safely because of its heat and radiation output.

Thermal, chemical and radiological gradients operate on the environment over periods as long as 500,000 years. Some of the concerns being expressed border on over-reaction to a problem that exists. It is not that one should wish away the problem.

But on the other hand, the reaction or concern is often inflated. As Tanner asked "Are not we kidding ourselves when we claim to be so concerned about the far-out possibility that a nuclear-waste-disposal site may begin to leak 10,000 or 1,000,000 years from now? In what other area of life do we show such foresight?"

We are confronted with a dilemma. On one side, 50–100 years hence, our fossil fuel sources may be reaching the rock-bottom of availability and the renewable sources of energy (solar, wind, geothermal, etc., power sources) may not meet the demands of society. Till alternate energy sources are developed, the only source available to mankind is the nuclear power. To set the scenario in proper perspective, it should be noted that nuclear power plants are managed subject to several radiation protection control practices.

Secondly, one may also note that "a 1000 MW electric coal-fired power plant releases into the environment nearly 6 million tons of greenhouse gases, 500,000 tons of mixtures of sulphur and nitrogen oxides and about 320,000 tons of ashes". These ashes containing NORMs are potentially capable of subjecting humanity to a collective dose of radiation higher than that attributable to wastes discharged into the environment by nuclear power plants generating the same amount of electricity. In spite of this ground reality, public perception about nuclear wastes is rather skewed against nuclear power in several countries.

Quantifying Natural and Artificial Nuclear Waste

The level of radioactive waste is quoted in terms of volume (in cubic metres) or in tonnage. Another way is to quote the radioactivity contained in such waste in bequerels (Bq). Both the units are useful because one needs to know the volume or weight of the waste to be handled for disposal purposes and also the radioactivity contained therein. We have already noted that nuclear waste from natural sources, including mining and related operations, could have resulted in production of radioactive waste of a few EBq and the sea is repository of several thousand EBq of radioactivity.

Compared to this it is estimated that in the military nuclear operations, the cold-war era resulted in release of more than 1000 EBq of nuclear debris in the atmosphere. Production of weapon-grade material resulted in about 1000 EBq of residual waste and "accidents and losses" of nuclear submarines and nuclear-powered satellites might have resulted in waste of a few EBq.

In the civilian regime, it is estimated that the nuclear waste, as a result of nuclear power production around the world over the past 50 years, is of the order of 1000 EBq and is growing at the rate of approximately 100 EBq/year.

Typically, a large nuclear power plant of generating capacity of 1000 MW electricity produces "around 27 tonnes of high-level radioactive waste, 310 tonnes of intermediate-level and 460 tonnes of lowlevel radioactive waste".

CLASSIFICATION OF RADIOACTIVE WASTE

Nuclear waste can be generally classified as either "lowlevel" radioactive waste or "high-level" radioactive waste.

Low-level Radioactive Waste

Basically all radioactive waste that is not high-level radioactive waste or intermediate-level waste or transuranic waste is classified as low-level radioactive waste. Volume-wise it may be larger than that of highlevel radioactive waste or intermediate-level radioactive waste or transuranic waste, but the radioactivity contained in the low-level radioactive waste is significantly less and made up of isotopes having much shorter halflives than most of the isotopes in high-level radioactive waste or intermediate-level waste or transuranic waste.

Large amounts of waste contaminated with small amounts of radionuclides, such as contaminated equipment (glove boxes, air filters, shielding materials and laboratory equipment) protective clothing, cleaning rags, etc., constitute low-level radioactive waste. Even components of decommissioned reactors may come under this category (after part decontamination procedures).

The level of radioactivity and half-lives of radioactive isotopes in low-level waste are relatively small. Storing the waste for a period of 10 to 50 years will allow most of the radioactive isotopes in low-level waste to decay, at which point the waste can be disposed of as normal refuse.

It may come as a surprise that several investigations have shown that exposure of mammals to low levels of radiation may indeed be beneficial, including, "increased life span, greater reproductive capacity, better disease resistance, increased growth rate, greater resistance to higher radiation doses, better neurological function, better wound healing and lower tumour induction and growth". Beneficial effects on plants include accelerated growth and development and increased harvests. Low-level radioactive waste, therefore, seems to be benign.

High-level Radioactive Waste

High-level radioactive waste is conceptualised as the waste consisting of the spent fuel, the liquid effluents arising from the reprocessing of spent fuel and the solids into which the liquid waste is converted. It consists, generally, material from the core of a nuclear reactor or a nuclear weapon. This waste includes uranium, plutonium and other highly radioactive elements created during fission, made up of fission fragments and transuranics.

These two components have different times to decay. The radioactive fission fragments decay to different stable elements via different nuclear reaction chains involving a, b and g emissions to innocuous levels of radioactivity, and this would take about 1000 years. On the other hand, transuranics take nearly 500,000 years to reach such levels. Heat output lasts over 200 years. Most of the radioactive isotopes in high-level waste emit large amounts of radiation and have extremely long half-lives (some longer than 100,000 years), creating long time-periods before the waste will settle to safe levels of radioactivity.

As a thumb-rule one may note that "volumes of lowlevel radioactive waste and intermediate-level waste greatly exceed those of spent fuel or high-level radioactive waste". In spite of this ground reality, the public concerns regarding disposal of high-level radioactive waste is worldwide and quite controversial.

APPROACHES TO RADIOACTIVE WASTE DISPOSAL

Waste disposal is discarding waste with no intention of retrieval. Waste management means the entire sequence of operations starting with generation of waste and ending with disposal.

Solid waste disposal, of waste such as municipal garbage, is based on three well-known methods, namely landfills, incineration and recycling. Sophisticated methods of landfills are adapted for radioactive waste also. However, during incineration of ordinary waste, fly ash, noxious gases and chemical contaminants are released into the air. If radioactive waste is treated in this manner, the emissions would contain radioactive particulate matter.

Hence when adapted, one uses fine particulate filters and the gaseous effluents are diluted and released. Recycling to some extent is feasible.

We have already dealt with the reprocessing approach, whereby useful radioactive elements are recovered for cyclic use. But it still leaves some waste that is a part of the high-level radioactive waste. Radioactive waste management involves minimising radioactive residues, handling waste-packing safely, storage and safe disposal in addition to keeping sites of origin of radioactivity clean.

Poor practices lead to future problems. Hence choice of sites where radioactivity is to be managed safely is equally important in addition to technical expertise and finance, to result in safe and environmentally sound solutions. The International Atomic Energy Agency (IAEA) is promoting acceptance of some basic tenets by all countries for radioactive waste management. These include:

- Securing acceptable level of protection of human health.
- Provision of an acceptable level of protection of environment.
- Assurance of negligible effects beyond national boundaries.
- Acceptable impact on future generations.
- No undue burden on future generations.

There are other legal, control, generation, safety and management aspects also. Next we review some approaches for radioactive waste disposal. To begin with, the radioactive waste management approach is to consider the nature of radioactive elements involved in terms of their half-lives and then choose the appropriate method of handling. If the concentrations of radioactive elements are largely shortlived, then one would resort to what is referred to as "delay and decay" approach; that is, to hold on to such a waste for a sufficiently long time that the radioactivity will die in the meanwhile.

A second approach is to "dilute and disperse" so that the hazard in the environment is minimized. But when the radioactivity is long-lived, the only approach that is possible is to "concentrate and contain" the activity. In order to carry out concentrating the waste (generally the sludge), chemical precipitation, ion exchange, reverse osmosis and natural or steam evaporation, centrifuging, etc. are resorted to.

The resulting solids are highly concentrated in radioactivity. The approaches that are being advocated or are currently in practice. However, to the extent that the mining operations result in "bringing the radioactivity to the surface and change its chemical and physical form that may increase its mobility in the environment", they assume importance in radioactive waste management.

Long-lived isotopes like 230Th, 226Ra, the decay products of uranium are part of the tailings and hence the tailings have to be contained. Low-level radioactive waste and even transuranic waste is often buried in shallow landfills.

One has to pay attention to any groundwater contamination that may result due to this. The highly radioactive liquid effluents are expected to be ultimately solidified into a leach-resistant form such as borosilicate glass, which is fairly robust in the sense that it is chemically durable, resistant to radiolysis, relatively insensitive to fluctuations in waste composition and easy to process remotely. (Immobilisation in cement matrices or bitumanisation or polymerisation are also some of the other options that are practised to some extent.) However, it must be noted that plutonium does not bind strongly to the matrix of the glass and "thus can be loaded only in trace amounts to prevent the possibility of criticality or recovery for clandestine purposes".

This glass in turn is placed in canisters made of specific alloys. Choice of the canister material would depend on the ultimate site where the waste will be disposed-off.

For example, if the ultimate disposal is in the oceans, the alloy chosen must have low corrosion rates under the environmental temperature, pressure, oxygen concentration, etc. Studies have been carried out in this respect. For example, it is found that in oxygenated sea water at 250°C, 7 mega Pascals pressure and 1750 ppm of dissolved oxygen, the corrosion rates of 1018 mild steel, copper, lead, 50: 10 cupro-nickel, Inconel 600 and Ticode 12 are 11.0, 5.0, 1.0, 0.7, 0.1 and 0.06 mm/year, respectively.

One seeks to dispose-off the high-level radioactive waste packages contained in multiple metal-barrier canisters within natural or man-made barriers, to contain radioactivity for periods as long as 10,000 to 100,000 years.

"The barrier is a mechanism or medium by which the movement of emplaced radioactive materials is stopped or retarded significantly or access to the radioactive materials is restricted or prevented". It is obvious that recourse to multiple barriers may assure safety of emplaced radioactivity over long periods of time.

The man-made barriers, namely the form to which waste is reduced, for example, in the glassy form, and the canister along with overpackaging, go along with natural barriers. As far as the choice of natural barriers is concerned, land-based mined depositories over fairly stable geologic formations are preferred over disposal in the oceans. However, several social and environmental concerns have prevented the land-route being adopted in counties like USA even after 50 years of accumulation of radioactive waste.

Therefore proposals have been made to take to the ocean-route and there also the choice varies from just placement of the canisters over the seabed to placement within the sub-seabed sediments and even within the basement rocks. In the US, as spent fuels have reached levels of radioactivity of the order of 50,000 MCi (excluding military sources), there is dearth of space to store additional irradiated fuel removed from operating reactors. Legally, the Department of Energy (DOE) is expected to take charge of all commercial spent fuel. However, the DOE has run into a dead-end.

On one hand it is unable to use spent fuel and on the other, its attempts to develop a permanent repository at Yucca Mountain in Nevada are met by social and State challenges as well as lack of complete study of the site itself. Presidential consent has not been forthcoming to any legislation in this connection.

OPTIONS BEING AIRED FOR DISPOSING RADIOACTIVITY

Triet Nguyen, Department of Nuclear Engineering, University of California, Berkeley, has written in an article "High-level Nuclear Waste Disposal", 14th November 1994 that "High-level nuclear waste from both commercial reactors and defence industry presents a difficult problem to the scientific community. The solutions to this problem are still debatable, both technically and ethically.There are many proposals for disposing high-level nuclear wastes.

However, the most favoured solution for the disposal of these wastes is isolating radioactive waste from man and biosphere for a period of time such that any possible subsequent release of radionuclides from the waste repository will not result in undue radiation exposure.

The basic idea behind this is to use stable geological environments that have retained their integrity for millions of years to provide a suitable isolation capacity for the long time-periods required".

The reason for relying on such geological environments is based on the following main consideration: "Geological media is an entirely passive disposal system with no requirement for continuing human involvement for its safety.

It can be abandoned after closure with no need for continuing surveillance or monitoring. The safety of the system is based on multiple barriers, both engineered and natural, the main one being the geological barrier itself". One way of disposing high-level nuclear waste materials which meets the above condition is the concept of disposing of these wastes by burial in suitable geologic media beneath the deep ocean floor, which is called seabed disposal.

The following options have been aired sometime or the other. Each one of the options demands serious studies and technical assessments:

- Deep geological repositories.
- Ocean dumping.
- Seabed burial.
- Sub-seabed disposal.
- Subductive waste disposal method.
- Transforming radioactive waste to non-radioactive stable waste.

- Dispatching to the Sun.

Major problems due to legal, social, political and financial reasons have arisen in execution due to:

- Environmental perceptions.
- Lack of awareness and education.
- "Not-in-my-backyard" syndrome.
- "Not-in-the-ocean" syndrome.
- Lack of proven technology.

Geologic Disposal

Geologic disposal in deep geological formations – whether under continental crust or under seabed – as a means of radioactive waste disposal has been recognised since 1957, for handling long-lived waste. Quite often, contrary to views expressed by environmentalists, it is 'not chosen as a cheap and dirty option to get the radioactive waste simply 'out of site and out of mind'.

The deep geological sites provide a natural isolation system that is stable over hundreds of thousands of years to contain long-lived radioactive waste. In practice it is noted that low-level radioactive waste is generally disposed in near-surface facilities or old mines. High-level radioactive waste is disposed in host rocks that are crystalline (granitic, gneiss) or argillaceous (clays) or salty or tuff. Since, in most of the countries, there is not a big backlog of high-level radioactive waste urgently awaiting disposal, interim storage facilities, which allow cooling of the wastes over a few decades, are in place.

Ocean Dumping

For many years the industrialised countries of the world opted for the least expensive method for disposal of the wastes by dumping them into the oceans. Before 1982, when the United States Senate declared a moratorium on the dumping of radioactive wastes, the US dumped an estimated 112,000 drums at thirty different sites in the Atlantic and Pacific oceans. Though this practice has been banned by most of the countries with nuclear programmes, the problem still persists. Russia, which currently controls sixty per cent of the world's nuclear reactors, continues to dispose of its nuclear wastes into the oceans.

According to Russia's Minister of Ecology, it will continue to dump its wastes into the oceans because it has no other alternative method. It will continue to do so until it receives enough international aid to create proper storage facilities. In response, the United States has pledged money to help Russia, but the problem continues. Although radioactive waste has known negative effects on humans and other animals, no substantial scientific proof of bad effects on the ocean and marine life has been found.

Hence some nations have argued that ocean-dumping should be continued. Others argue that the practice should be banned until further proof of no harm is available. Oceanic Disposal Management Inc., a British Virgin Islands company, has also proposed disposing of nuclear and asbestos waste by means of Free-Fall Penetrators.

Essentially, waste-filled missiles, which when dropped through 4,000 m of water, will embed themselves 60–80 m into the seabed's clay sediments. These penetrators are expected to survive for 700 to 1,500 years. Thereafter the waste will diffuse through the sediments. This was a method considered by the Scientific Working Group (SWG) of the Nuclear Energy Agency (NEA) during the eighties.

Penetrator disposal is potentially both feasible and safe, its implementation would depend on international acceptance and the development of an appropriate international regulatory framework. Neither of these exists, nor are they likely to in the foreseeable future. The penetrator method has also been further constrained by a recent revision of the definition of "dumping", by the London Dumping Convention, to include "any deliberate disposal or storage of wastes or other matter in the seabed and the subsoil thereof".

Sub-Seabed Disposal

Seabed disposal is different from sea-dumping which does not involve isolation of low-level radioactive waste within a geological strata. The floor of deep oceans is a part of a large tectonic plate situated some 5 km below the sea surface, covered by hundreds of metres of thick sedimentary soft clay.

These regions are desert-like, supporting virtually no life. The Seabed Burial Proposal envisages drilling these 'mud-flats' to depths of the order of hundreds of metres, such boreholes being spaced apart several hundreds of metres.

The high-level radioactive waste contained in canisters, to which we have referred to earlier, would be lowered into these holes and stacked vertically one above the other interspersed by 20 m or more of mud pumped in. The proposal to use basement-rock in oceans for radioactive waste disposal is met with some problems: variability of the rock and high local permeability. Oceanic water has a mixing time of the order of a few thousand years which does not serve as a good barrier for long-lived radionuclides.

Subductive Waste Disposal Method

This method is the state-of-the-art in nuclear waste disposal technology. It is the single viable means of disposing radioactive waste that ensures non-return of the relegated material to the biosphere. At the same time, it affords inaccessibility to eliminated weapons material. The principle involved is the removal of the material from the biosphere faster than it can return. It is considered that "the safest, the most sensible, the most economical, the most stable long-term, the most environmentally benign, the most utterly obvious places to get rid of nuclear waste, high-level waste or lowlevel waste is in the deep oceans that cover 70% of the planet".

Subduction is a process whereby one tectonic plate slides beneath another and is eventually reabsorbed into the mantle. The subductive waste disposal method forms a high-level radioactive waste repository in a subducting plate, so that the waste will be carried beneath the Earth's crust where it will be diluted and dispersed through the mantle. The rate of subduction of a plate in one of the world's slowest subduction zones is 2.1 cm annually.

This is faster than the rate (1 mm annually) of diffusion of radionuclides through the turbidite sediments that would overlay a repository constructed in accordance with this method. The subducting plate is naturally predestined for consumption in the Earth's mantle. The subducting plate is constantly renewed at its originating oceanic ridge. The slow movement of the plate would seal any vertical fractures over a repository at the interface between the subducting plate and the overriding plate.

Transmutation of High-level Radioactive Waste

This route of high-level radioactive waste envisages that one may use transmutational devices, consisting of a hybrid of a subcritical nuclear reactor and an accelerator of charged particles to "destroy" radioactivity by neutrons. "Destroy" may not be the proper word; what is effected is that the fission fragments can be transmuted by neutron capture and beta decay, to produce stable nuclides.

Transmutation of actinides involves several competing processes, namely neutron-induced fission, neutron capture and radioactive decay. The large number of neutrons produced in the spallation reaction by the accelerator are used for "destroying" the radioactive material kept in the subcritical reactor. The scheme has not yet been demonstrated to be practical and costeffective.

Solar Option

It is proposed that "surplus weapons" plutonium and other highly concentrated waste might be placed in the Earth orbit and then accelerated so that waste would drop into the Sun. Although theoretically possible, it involves vast technical development and extremely high cost compared to other means of waste disposal. Robust containment would be required to ensure that no waste would be released in the event of failure of the "space transport system".

Other Options and Issues

In its 1994 report entitled "Management and Disposition of Excess Weapons Plutonium", the National Academy of Sciences set forth two standards for managing the risks associated with surplus weapons-usable fissile materials. First, the storage of weapons should not be extended indefinitely because of non-proliferation risks and the negative impact it would have on armsreduction objectives.

Second, options for long-term disposition of plutonium should seek to meet a "spent-fuel standard" in which the plutonium is made inaccessible for weapons use. One of the chosen options of DOE is for dealing with surplus plutonium, its use as a Mixed Oxide Fuel (MOX) to be burned in reactors such as the CANDU. The United States policy is not to encourage the civil use of plutonium. The Nuclear Control Institute regards the vitrification approach as posing fewer risks than the MOX approach with regard to diversion or theft of warhead material, reversal of the disarmament process, and other adverse effects on international arms control and non-proliferation efforts.

A decision to dispose-off warhead plutonium by means of vitrification or other immobilisation technology would be an essential step toward achievement of such a regime. Proponents of MOX disposition claim that vitrification technology is immature, speculative and cannot be ready soon enough.

On the other hand, the MOX option, though it does not necessarily involve further reprocessing, would clearly encourage civilian use of plutonium, which in some countries like Japan even includes plans for reprocessing irradiated MOX fuel. In the opinion of the Nuclear Control Institute, "the MOX option" sends the wrong signal in three ways.

First, this option effectively declares that plutonium has an asset value, and that the energy contained within it should be viewed as a "national asset" (as the US DOE expressed it) or even "national treasure" when, in fact, plutonium fuel has been shown to be an economic liability. Second, the MOX option suggests that a commercial plutonium fuel cycle can be effectively safeguarded, when, in fact, it is becoming obvious that large-throughput plutonium plants face daunting safeguard problems.

Third, the MOX option would be portrayed as giving credibility to the claim that plutonium recycle in light water reactors (LWRs) is essential to nuclear waste management, at a time when direct disposal of spent fuel is looking increasingly attractive to utilities. There are other arguments that relate to proliferation using high-level radioactive waste. It is believed that the technologies of Laser Isotope Separation and the Large Volume

Plasma Process may permit the mining of weapons materials from any matrix.

There are many international transporting-related issues. It is not uncommon that reprocessing of one country's spent fuel or waste is taken up in a different country. Such movement is often via one or more countries or over the international waters. Regulatory mechanisms, both national and international, have to be in place to guarantee safety of the waste under these conditions.

RADIOACTIVE WASTE MANAGEMENT IN INDIA

Just as per capita consumption of electricity is related to the standard of living in a country, the electricity generation by nuclear means can be regarded as a minimum measure of radioactive waste that is generated by a country and hence the related magnitude of radioactive waste management.

On the scale of nuclear share of electricity generation, India ranks fourth from the bottom in about 30 countries. As of the year 2000, India's share of nuclear electricity generation in the total electricity generation in the country was 2.65% compared to 75%, 47%, 42.24%, 34.65%, 31.21%, 28.87%, 19.80%, 14.41% and 12.44% of France, Sweden, the Republic of Korea, Japan, Germany, UK, USA, Russia and Canada, respectively.

The reactors in operation produce in net Gigawatts (one billion (109) watts) in the latter countries nearly 63, 9,13, 44, 21, 13, 97, 20 and 10, respectively; India's reactors in operation yield 1.9 on this scale (both data are as per IAEA Report of 2000).

Hence the magnitude of radioactive waste management in India could be miniscule compared to that in other countries, especially when one takes into account the nuclear arsenal already in stockpile in the nuclear weapons countries. As more power reactors come onstream and as weaponisation takes deeper routes the needs of radioactive waste management increase and in this context the experience of other countries would provide useful lessons.

Radioactive waste management has been an integral part of the entire nuclear fuel cycle in India. Low-level radioactive waste and intermediate-level waste arise from operations of reactors and fuel reprocessing facilities. The low-level radioactive waste liquid is retained as sludge after chemical treatment, resulting in decontamination factors ranging from 10 to 1,000. Solid radioactive waste is compacted, bailed or incinerated depending upon the nature of the waste.

Solar evaporation of liquid waste, reverse osmosis and immobilisation using cement matrix are adopted depending on the form of waste. Underground engineered trenches in near-surface disposal facilities are utilised for disposal of solid waste; these disposal sites are under continuous surveillance and monitoring. High efficiency particulate air (HEPA) filters are used to minimise air-borne radioactivity. Over the past four decades radioactive waste management facilities have been set up at Trombay, Tarapore, Rawatbhata, Kalpakkam, Narora, Kakrapara, Hyderabad and Jaduguda, along with the growth of nuclear power and fuel-reprocessing plants. Multiplebarrier approach is followed in handling solid waste.

After the commissioning of the fast breeder test reactor at Kalpakkam, one is required to reprocess the burnt carbide fuel from this reactor. As the burn-up of this fuel is likely to be of the order of 100 MWD/kg, nearly an order of magnitude more than that of thermal reactors and due to short cooling-time before reprocessing, specific activity to be handled will be greatly enhanced. The use of carbide fuel would result in new forms of chemicals in the reprocessing cycle. These provide new challenges for fast-reactor fuel reprocessing.

RADIOACTIVE WASTE DISPOSAL: AN ENVIRONMENTAL PERSPECTIVE

Any activity that produces or uses radioactive materials generates radioactive waste. Mining, nuclear power generation, and various processes in industry, defence, medicine, and scientific research produce byproducts that include radioactive waste. Radioactive waste can be in gas, liquid or solid form, and its level of radioactivity can vary. The waste can remain radioactive for a few hours or several months or even hundreds of thousands of years.

Because it can be so hazardous and can remain radioactive for so long, finding suitable disposal facilities for radioactive waste is difficult. Depending on the type of waste disposed, the disposal facility may need to contain radiation for a very long time. Proper disposal is essential to ensure protection of the health and safety of the public and quality of the environment including air, soil, and water supplies.

Radioactive waste disposal practices have changed substantially over the last twenty years. Evolving environmental protection considerations have provided the impetus to improve disposal technologies, and, in some cases, clean up facilities that are no longer in use. Designs for new disposal facilities and disposal methods must meet environmental protection and pollution prevention standards that are more strict than were foreseen at the beginning of the atomic age.

Disposal of radioactive waste is a complex issue, not only because of the nature of the waste, but also because of the complicated regulatory structure for dealing with radioactive waste. There are a variety of stakeholders affected, and there are a number of regulatory entities involved. Federal government agencies involved in radioactive waste management include: the Environmental Protection Agency (EPA), the Nuclear Regulatory Commission (NRC), the Department of Energy (DOE), and the Department of Transportation. In addition, the states and affected Indian Tribes play a prominent role in protecting the public against the hazards of radioactive waste.

Types of Radioactive Waste

There are six general categories of radioactive waste:

- Spent nuclear fuel from nuclear reactors.
- High-level waste from the reprocessing of spent nuclear fuel.
- Transuranic waste mainly from defence programmes.
- Uranium mill tailings from the mining and milling of uranium ore.
- Low-level waste.
- Naturally occurring and accelerator-produced radioactive materials.

Radioactive waste is categorised according to its origin and not necessarily according to its level of radioactivity. For example, some low-level waste has the same level of radioactivity as some high-level waste.

SPENT NUCLEAR FUEL AND HIGH-LEVEL RADIOACTIVE WASTE

In addition to being used to generate commercial electricity, nuclear reactors are used in government-sponsored research and development programs, universities and industry; in science and engineering experimental programmes; at nuclear weapons production facilities; and by the US Navy and military services. The operation of nuclear reactors results in spent reactor fuel. The reprocessing of that spent fuel produces high-level radioactive waste (HLW).

The fuel for most nuclear reactors consists of pellets of ceramic uranium dioxide that are sealed in hundreds of metal rods. These rods are bundled together to form what is known as a "fuel assembly". Depending upon the type and size of the reactor, a fuel assembly can weigh up to 1,500 pounds.

As the nuclear reactor operates, uranium atoms fission (split apart) and release energy. When most of the usable uranium has fissioned, the "spent" fuel assembly is removed from the reactor. Until a disposal or long-term storage facility is operational, most spent fuel is stored in water pools at the reactor site where it was produced. The water removes leftover heat generated by the spent fuel and serves as a radiation shield to protect workers at the site.

The operation of nuclear reactors over the last twenty years has substantially added to the amount of radioactive waste in this country. HLW is the liquid waste that results when spent fuel is reprocessed to recover unfissioned uranium and plutonium. During this process, the fuel is dissolved by strong chemicals, and this results in liquid HLW. Plans are to solidify these liquids into a form that is suitable for disposal. Solidification is still in the planning stages.

While currently there are no commercial facilities in this country that reprocess spent fuel, spent fuel from defence programme reactors has been routinely reprocessed for use in producing nuclear weapons or for reuse in new fuel.

Compared to the total inventory of HLW, the volume of commercial HLW from the reprocessing of commercial spent fuel is almost insignificant, less than one per cent. Defence-related HLW comprises greater than ninety-nine per cent of the volume of HLW.

HLW is now stored in underground tanks or stainless steel silos on federal reservations in South Carolina, Idaho, and Washington and at the Nuclear Fuel Services Plant in West Valley, NY. These facilities have begun programs to solidify and structurally stabilise the waste in preparation for disposal at a national repository.

Regulation of Disposal

Some elements, such as plutonium, in HLW and spent fuel are highly radioactive and remain so for thousands of years. Therefore, the safe disposal of this waste is one of the most controversial environmental subjects facing the federal government and affected states.

The federal government (the EPA, the DOE, and the NRC) has overall responsibility for the safe disposal of HLW and spent fuel. The EPA is responsible for developing environmental standards that apply to both DOE-operated and NRC-licensed facilities. Currently, the NRC is responsible for licensing such facilities and ensuring their compliance with the EPA standards.

DOE is responsible for developing the deep geologic repository which has been authorized by Congress for disposing of spent fuel and high level waste. Both the NRC and the Department of Transportation are responsible for regulating the transportation of these wastes to storage and disposal sites.

Site Selection for Storage and Disposal

In the early 1980's, the DOE formally adopted a national strategy to develop mined geologic repositories as disposal facilities for spent fuel and high-level radioactive waste. In 1983, the DOE identified nine potentially acceptable sites and, in 1984, selected three sites as candidates for further characterisation. In 1987, Congress directed the DOE to pursue the investigation of only the Yucca Mountain, NV site in order to determine whether the site is suitable for development as a repository. The DOE has designed a comprehensive "site characterisation" programme to evaluate the suitability of the Yucca Mountain site. The objectives of this programme are to: (1) determine the geologic, hydrologic, and geochemical conditions at Yucca Mountain; (2) provide information needed to design a package for the disposal of radioactive waste; (3) provide information for the design of the repository facility; and (4) evaluate whether Yucca Mountain can meet NRC and EPA protection and safety requirements.

The DOE is also developing plans for the siting and development of a potential Monitored Retrievable Storage (MRS) facility. The MRS facility could be used to receive and store spent fuel from commercial power reactors for subsequent shipment to a repository when such a facility becomes operational.

Setting Environmental Protection Standards

The management and disposal of spent nuclear fuel, HLW, and transuranic (TRU) wastes. The disposal portion of these standards was successfully challenged in the courts and returned to the Agency for revision. The court was primarily concerned that the regulations might not adequately protect ground water and individuals from radioactive contamination. Following the court's ruling in 1987, the EPA worked to re-promulgate the disposal portion of these standards.

In October 1992, two laws were enacted, the Waste Isolation Pilot Plant (WIPP) Land Withdrawal Act and the Energy Policy Act, that affected EPA's development of standards for the management and disposal of spent nuclear fuel, HLW and TRU wastes on TRU waste, EPA's Administrator issued the revised disposal standards as mandated by the WIPP Land Withdrawal Act in December 1993. These standards apply to all HLW, spent fuel, and TRU waste disposal except for disposal at the Yucca Mountain site. The Energy Policy Act directs the EPA to issue environmental standards, which protect public health and safety and are specific to the Yucca Mountain site.

The Act also requires that the National Academy of Sciences (NAS) conduct a study to provide findings and recommendations related to the form and content of environmental radiation protection standards for Yucca Mountain, Nevada. The EPA's standards for Yucca Mountain must be developed based upon the findings and recommendations of the NAS and must be issued within one year from the time the EPA receives the NAS recommendations. NRC, as the licensing authority for this site, must incorporate the EPA's environmental standards in their overall licensing regulations for HLW disposal (10 CFR 60).

TRANSURANIC RADIOACTIVE WASTE

Transuranic (TRU) waste materials have been generated in the US since the 1940s. Most of this waste originates from nuclear weapons production facilities for defence programmes. "Transuranic" refers to atoms of man-made elements that are heavier (higher in atomic number) than uranium.

The most prominent element in most TRU waste is plutonium. Some TRU waste consists of items such as rags, tools, and laboratory equipment contaminated with radioactive materials. Other forms of TRU waste include organic and inorganic residues or even entire enclosed contaminated cases in which radioactive materials were handled. Some TRU waste emits high levels of penetrating radiation; this type requires protective shielding. However, most TRU waste does not emit high levels of penetrating radiation but poses a danger when small particles of it are inhaled or ingested.

The radiation from the particles is damaging to lung tissue and internal organs. As long as this type of TRU waste remains enclosed and contained, it can be handled safely. Another problem with TRU waste is that most of its radioactive elements are long-lived. That is, they stay radioactive for a long time. For example, half of the original amount of plutonium-239 in the waste will remain harmful after 24,000 years. Disposal must be carefully planned so that the waste poses no undue threat to public health or the environment for years to come.The total volume of TRU waste and TRU contaminated soil is estimated at around one million cubic meters.

Site Selection for Storage and Disposal

In the past, much of the TRU waste was disposed of similarly to low-level radioactive waste, i.e., in pits and trenches covered with soil. In 1970, the Atomic Energy Commission (predecessor to the DOE) decided that TRU waste should be stored for easy retrieval to await disposal at a repository. Federal facilities in Washington, Idaho, California, Colorado, New Mexico, Nevada, Tennessee, South Carolina, Ohio, and Illinois are currently storing TRU waste.

The DOE has evaluated several alternatives for managing buried waste and contaminated soil including:

- Leaving it in place and monitoring it;
- Leaving it in place and improving the containment; and
- Removing, processing, and disposing of the waste in a repository.

As a first step in developing a permanent disposal site for TRU waste, the DOE is developing an underground, geologic repository called the

Waste Isolation Pilot Plant (WIPP), near Carlsbad, NM. This site has been excavated in a salt bed about 2,100 feet underground. The WIPP will have to meet environmental standards established by the EPA before it can be used as a permanent disposal site. If the WIPP site is eventually determined to be suitable for the disposal of TRU waste, the underground disposal area is planned to cover 100 acres. It will have a design capacity of over 2 million cubic metres, or about 850,000 barrels, of TRU waste.

Setting Environmental Protection Standards

As stated earlier, the EPA established environmental standards applicable to spent fuel, HLW and TRU waste, but they were returned to the Agency by the courts for revision. While the Energy Policy Act specifies procedures for developing standards for a repository at Yucca Mountain, NV, the Waste Isolation Pilot Plant (WIPP) Land Withdrawal Act requires the EPA to promulgate final standards applicable to WIPP and all other spent nuclear fuel, HLW, and TRU waste disposal facilities other than those developed under the Nuclear Waste Policy Act of 1982.

The WIPP Land Withdrawal Act reinstated all of the EPA's 1985 radioactive waste disposal standards except for the sections that the court found problematic, i.e., the Individual and Ground-Water Protection Requirements of the disposal standards. The reinstated sections consist primarily of containment requirements and assurance requirements. These requirements are designed to help ensure that the wastes will be disposed of in a manner that limits the release of radioactive materials.

In 1993, EPA finalised amendments to the standards to address the court's concerns. Individual radiation protection standards will limit a person's total annual radiation exposure, considering the sum of all possible exposures. Ground-water protection standards protect present and future sources of drinking water.

New Regulatory Responsibilities for EPA

Under the WIPP Land Withdrawal Act, Congress gave EPA the responsibility for implementing its radioactive waste disposal standards at the WIPP. The Act also requires the EPA to review and approve of the DOE's plans for testing and retrieving waste at the WIPP.

EPA must also ensure compliance with all federal environmental laws and regulations. In order for the WIPP to become a permanent disposal facility, the EPA must certify that the facility complies with its disposal standards. If the EPA does not certify the WIPP, the DOE must decommission the facility.

Even if the EPA certifies the WIPP, the Agency will have to determine, on an ongoing basis, whether it continues to comply with the disposal standards as well as all other federal environmental laws, regulations, and permit requirements that apply.

In particular, DOE must demonstrate that the WIPP complies with the Clean Air Act; the Comprehensive Environmental Response, Compensation, and Liability Act; the Solid Waste Disposal Act; the Safe Drinking Water Act; and the Resource Conservation and Recovery Act.

URANIUM MILL TAILINGS

Uranium mill tailings are the radioactive sandlike materials that remain after uranium is extracted by milling ore mined from the earth. Tailings are placed in huge mounds called tailings piles which are located close to the mills where the ore is processed. The most important radioactive component of uranium mill tailings is radium, which decays to produce radon. Other potentially hazardous substances in the tailings are selenium, molybdenum, uranium, and thorium.

Uranium mill tailings can adversely affect public health. There are four principal ways (or exposure pathways) that the public can be exposed to the hazards from this waste. The first is the diffusion of radon gas directly into indoor air if tailings are misused as a construction material or for backfill around buildings. When people breathe the air containing radon, it increases their risk of developing lung cancer.

Second, radon gas can diffuse from the piles into the atmosphere where it can be inhaled and small particles can be blown from the piles where they can be inhaled or ingested. Third, many of the radioactive decay products in tailings produce gamma radiation, which poses a health hazard to people in the immediate vicinity of tailings. Finally, the dispersal of tailings by wind or water, or by leaching, can carry radioactive and other toxic materials to surface or ground water that may be used for drinking water.

The NRC and some individual states that have regulatory agreements with the NRC have licensed 26 sites for milling uranium ore. However, most of the mills at these sites are no longer processing ore. Another 24 sites have been abandoned and are currently the responsibility of DOE.

All the tailings piles except for one abandoned site located in Canonsburg, PA, are located in the West, predominantly in arid areas. The licensed tailings piles contain a combined total of approximately 200 million metric tons (MT), with individual piles ranging from about 2 million MT to about 30 million MT. (A metric ton is 2,200 pounds.) The 24 abandoned sites contain a total of about 26 million MT and range in size from about 50 thousand MT to about 3 million MT.

It is unlikely that there will be much additional accumulation of mill tailings in the US, because foreign countries now produce uranium much more cheaply than can domestic producers.

Setting Environmental Protection Standards

The EPA issued two sets of standards controlling hazards from uranium mill tailings in 1983, under the authority of the Uranium Mill Tailings Radiation Control Act of 1978. These standards provide for the cleanup and disposal of mill tailings at abandoned sites and the disposal of tailings at licensed sites after cessation of operations.

They are implemented by DOE, NRC, and some states through agreements with NRC, and require a combination of active and passive controls to clean up contaminated ground water as well as tailings that have been misused at off-site locations, and to dispose of tailings in a manner that will prevent misuse, limit radon emissions, and protect ground water.

Active controls include building fences, putting up warning signs, and establishing land use restrictions. Passive controls include constructing thick earthen covers, protected by rock and designed to prevent seepage into ground water, over the waste. Earthen covers also effectively limit radon emissions and gamma radiation and, in conjunction with the rock covers, serve to stabilise the piles to prevent dispersion of the tailings through erosion or intrusion. In some cases, piles may be moved to safer locations.

The standards were amended in 1993 to require that all licensed sites that have ceased operation undergo remedial action as soon as possible. The EPA is in the process of enacting revised ground-water protection standards that will require the same treatment of ground water at the abandoned sites as is now required at the licensed sites.

In addition, EPA enacted Clean Air Act standards in 1989 limiting radon emissions and restricting the length of time that abandoned piles may remain uncovered with no controls on radon emissions.

EPA also requires that any piles that may be constructed in the future meet requirements that limit radon emissions and inhibit ground-water contamination during their operational phase. Licensed mills also are subject to the Uranium Fuel Cycle standard which regulates radionuclide emissions other than radon.

LOW-LEVEL RADIOACTIVE WASTE

Low-level radioactive waste (LLW) is radioactively contaminated industrial or research waste such as paper, rags, plastic bags, protective clothing, cardboard, packaging material, organic fluids, and water-treatment residues.

LLW is generated by government facilities, utilities, industries, and institutional facilities. In addition to 35 major DOE facilities, over 20,000 commercial users of radioactive materials generate some amount of LLW.

LLW generators include approximately 100 operating nuclear power reactors, associated fuel fabrication facilities, and uranium fuel conversion plants, which together are known as nuclear fuel-cycle facilities. Hospitals, medical schools, universities, radiochemical and radiopharmaceutical manufacturers and research laboratories are other users of radioactive materials which produce LLW.

The clean-up of contaminated buildings and sites will generate more LLW in the future. It also projects that the volume will double by 2020. The volume of low-level radioactive waste disposed of by major sources in the United States. Both commercial and defence-related LLW have been disposed of using shallow land disposal methods.

Disposal Management

The EPA has the authority to set generally applicable environmental standards for LLW disposal; such standards would be implemented by the NRC and the DOE. DOE is planning the clean-up of radioactively contaminated sites which will result in considerable volumes of LLW. Because of this, EPA is developing cleanup regulations as well as general environmental standards for LLW disposal. EPA plans to propose the disposal standards at the end of 1994. The standards will facilitate planning and reduce costs for clean-up and disposal.

The NRC and some individual states that have regulatory agreements with NRC regulate all disposal of commercial LLW. In 1982, the NRC improved its regulatory requirements. That year, NRC established disposal site performance objectives for land disposal of LLW technical requirements for the siting, design, operation, and closure for near-surface disposal facilities; technical requirements concerning waste packaging for land disposal; classification of waste; institutional requirements; and administrative and procedural requirements for licensing a disposal facility.

Though the 1982 NRC regulations exempted existing NRC disposal site licensees, NRC and the states are working to incorporate such requirements into those licenses. In 1988, the DOE, which is self-regulating, issued its own orders governing the DOE disposal sites.

The general regulatory framework for the disposal of LLW has changed to account for new technology, what we have learned from past disposal practices, and current wisdom about environmental protection. As a result of increasing costs of LLW disposal at existing sites, pre-disposal waste processing (e.g., volume reduction) is a more common practice.

The waste is processed by separating radioactive from nonradioactive components and by compacting bulk waste before packaging for disposal. Consequently, while the volume of waste to be disposed of is reduced, the concentration of radioactivity is greater. This waste requires more stringent safeguards for its disposal.

Site Selection for Disposal

The first of six regional, commercial LLW disposal sites was licensed in 1962. Since then, four of the commercial sites have closed, mainly because of problems with site instability. These problems included the collapse of the earth covering the waste and difficulties in managing surface- and ground-water contamination.

Since then the technology and requirements governing disposal sites have been upgraded. New disposal facilities must be designed to avoid two kinds of failures: those caused by long-term processes such as subsidence and those caused by more unpredictable events such as human intrusion (either intentional or unintentional) and natural disaster.

The Low-Level Radioactive Waste Policy Act of 1980 and subsequent amendments direct states to take care of their own LLW either individually or through regional groupings, referred to as compacts. The states are now in the process of selecting new LLW disposal sites to take care of their own waste. The selection process for these new sites is complex and varies because of many factors including the regulations for site selection. This selection process will be affected by EPA's new LLW standard.

DISPOSAL OF NATURALLY OCCURRING AND ACCELERATOR-PRODUCED WASTES

Accelerator-produced Materials

Accelerator-produced radioactive waste is produced during the operation of atomic particle accelerators for medical, research, or industrial purposes. The accelerators use magnetic fields to move atomic particles at higher and higher speeds before crashing into a preselected target. This reaction produces desired radioactive materials in metallic targets or kills cancer cells where a cancer tumor is the target. The radioactivity contained in the waste from accelerators is generally short-lived, less than one year.

The waste may be stored at laboratories or production facilities until it is no longer radioactive. An extremely small fraction of the waste may retain some longer-lived radioactivity with half lives greater than one year. There are no firm estimates of the amount of this type of radioactive waste; however, it is generally accepted that the volume is extremely small compared to the other wastes.

Naturally Occurring Radioactive Materials (NORM)

Naturally occurring radioactive materials (NORM) generally contain radionuclides found in nature. Once NORM becomes concentrated through human activity, such as mineral extraction, it can become a radioactive waste. There are two types of naturally occurring radioactive waste: discrete and diffuse.

The first, discrete NORM, has a relatively high radioactivity concentration in a very small volume, such as a radium source used in medical procedures. Estimates of the volumes of discrete NORM waste are imprecise, and the EPA is conducting studies to provide a more accurate assessment of how much of this waste requires attention. Because of its relatively high concentration of radioactivity, this type of waste poses a direct radiation exposure hazard.

The second type, diffuse NORM, has a much lower concentration of radioactivity, but a high volume of waste. This type of waste poses a different type of disposal problem because of its high volume.

Chapter 9

Effect of Urea on Soil

Favourable economics of manufacturing, handling, storage, and transportation have made urea a very competitive source of fertilizer N. Worldwide urea use is almost five times that of NH_4NO_3—Urea is the principal form of dry fertilizer N in the United States, approaching 16% of total N use.

Granular urea has noteworthy characteristics, including less tendency to stick and cake than NH_4NO_3 lack of sensitivity to fire and explosion, and less corrosiveness to handling and application equipment. Substantial savings in handling, storage, transportation, and application costs are possible because of urea's high N content.

BIURET LEVELS

The concentration of biuret (NH_2-CO-NH-CO-NH_2) in urea is of special concern because of its phytotoxicity. Biuret levels of 2% can be tolerated in most fertilizer programmes. Because citrus, pineapple, and other crops are sensitive to biuret in urea applied as a foliar spray, less than 0.25% biuret is recommended. Solutions made from urea containing 1.5% biuret are acceptable for foliar application on corn and soybeans. Urea high in biuret should not be placed near or in the seed row.

BEHAVIOUR OF UREA IN SOILS

When applied to soil, urea is hydrolyzed by the enzyme urease to NH_4+. Depending on soil pH, the NH_4+ may form NH_3, which can be volatilised at the soil surface, as represented in the following reactions:

$$CO(NH_2)^2 + H + 2H_2O \rightarrow 2NH_4{+} + HCO_3$$

$$NH_4{+} \rightarrow NH_3 + H$$

Urea hydrolysis proceeds rapidly in warm, moist soils, with most of the urea transformed to NH_4^+ in several days. Urease, an enzyme that catalyzes the hydrolysis of urea, is abundant in soils. Large numbers of bacteria, fungi, and actinomycetes in soils possess urease. Urease activity increases with the size of the soil microbial population and with OM content. The presence of fresh plant residues often results in abundant supplies of urease. Urease activity is greatest in the rhizosphere, where microbial activity is high and where it can accumulate from plant roots. Rhizosphere urease activity varies depending on the plant species and the season of the year. Although temperatures up to 37°C favour urease activity, hydrolysis of urea occurs at temperatures down to 2°C and lower. This evidence of urease functioning at low temperatures, combined with urea's ability to melt ice at temperatures down to 1°C IF (–12°C), suggests that a portion of fall or early-winter-applied urea may be converted to NH_3 or NH_4 before the spring.

The effects of soil moisture on urease activity are generally small in comparison to the influence of temperature and pH. Hydrolysis rates are highest at soil moisture contents optimum for plants. Free NH_3 inhibits the enzymatic action of urease. Since significant concentrations of free NH_3 can occur at pH values above 7, some temporary inhibition of urease by free NH_3 occurs after the addition of urea because soil pH in the immediate vicinity of the urea source may reach values of up to 9.0. High rates of urea fertilization in localized placement could create conditions restrictive to the action of urease.

MANAGEMENT OF UREA FERTILIZER

Careful management of urea and urea-based fertilizers will reduce the potential for NH_3 volatilisation losses and increase the effectiveness of urea fertilizers. Surface applications of urea are most efficient when they are washed into the soil or applied to soils with low potential for volatilisation. Conditions for best performance of surface-applied urea are cold, dry soils at the time of application and/or the occurrence of significant precipitation, probably more than 0.25 cm (0. 1 inch), within the first 3 to 6 days following application.

Movement of soil moisture containing dissolved NH_3 and diffusion of moisture vapour to the soil surface during the drying process probably contribute to NH_3 volatilisation at or near the soil surface. Incorporation of broadcast urea into soil minimises NH_3 losses by increasing the volume of soil to retain NH_3. Also, NH_3 not converted in the soil must diffuse over much greater distances before reaching the atmosphere. If soil and other environmental conditions appear favorable for NH3 volatilisation, deep incorporation is preferred over shallow surface tillage.

Band placement of urea results in soil changes comparable to those produced by applications of anhydrous NH_3. Diffusion of urea from banded applications can be 2.5 cm (1 in.) within 2 days of its addition, while appreciable amounts of NH_4 can be observed at distances of 3.8 cm (1.5 in.) from the band. After dilution or dispersion of the band by moisture movement, hydrolysis begins within 3 to 4 days or less under favorable temperature conditions.

Placement of urea with the seed at planting should be carefully controlled be- cause of the toxic effects of free NH_3 on germinating seedlings. The harmful effects of urea placed in the seed row can be eliminated or greatly reduced by banding at least 2.5 cm (1 in.) directly below and/or to the side of the seed row of most crops.

Seed placed urea should not exceed 5 to 10 lbs N/a. The effect on germination of urea placed near seeds is influenced by available soil moisture. With adequate soil moisture in medium-textured loam soils at seeding time, urea at 30 lb N/a can be used without reducing germination and crop emergence. However, in low-moisture, coarse-textured (sandy loam) soils, urea at 10 to 20 lb N/a often reduces both germination and crop yields. Seedbed moisture is less critical in fine-textured (clay and clay loam) soils, and urea can usually be drilled in at rates of up to 30 lb N/a. To summarise, the effectiveness of urea depends on the interaction of many factors, which cause some variability in the crop response to urea. However, if managed properly, urea will be about as effective as the other N sources.

NITROGEN EFFICIENCY OF UREA FERTILIZERS

Certain nitrogen-supplying fertilizers are more prone to lose nitrogen (N) into the atmosphere through ammonia volatilisation than others. An AAES study examined the effect of different application methods and additives that may help make these fertilizers less volatile and found that these treatments had little effect on volatilisation.

Surface-applied urea and urea-ammonium nitrate solutions (AN) tend to be more volatile than ammonium nitrate. Ammonia volatilisation—the loss of free ammonia to the atmosphere—involves several steps. First, an enzyme (urease) present in soils and organic residue acts on urea and converts it to an unstable form. This unstable form can quickly change to ammonia and carbon dioxide. Under ideal conditions, the ammonia is converted to ammonium, ready for plant uptake. But, under less than ideal conditions, the ammonia can be lost to the atmosphere.

Many interrelating factors influence the amount of N lost through ammonia volatilisation. These factors include urease activity, temperature, soil moisture, application method, soil pH, and soil cation exchange capacity. The greatest losses will most likely occur when urea or urea-containing fertilizer is surface-applied to a soil where high amounts of plant residue are present.

Losses will likely be accelerated if soil is moist at application, followed by five to seven hot and windy dry days. Urease activity is higher on plant surfaces and organic residues than in soils. Urea hydrolyses rapidly when in contact with these residues, and conditions are created around the fertilizer that favour the formation of ammonia. This rapid hydrolysis of urea on the soil surface can cause the loss of ammonia to the atmosphere, especially when rapid soil drying conditions exist. Losses through ammonia volatilisation can be essentially eliminated when urea is injected into the soil at application or incorporated into soil within 48 hours after application either mechanically or by rainfall.

Band applications of urea-containing N solutions, such as UAN, will most likely result in less volatilisation losses than broadcast applications. Banding the solution will give more opportunity for applied urea to contact the soil surface instead of being held by organic residue, which contains high levels of urease. Frequently, UAN solutions may be used as a carrier for herbicides to control weeds in winter cereal crops. This saves a trip across the field and provides a more potent herbicide mix than does a herbicide/water mix, but does not fit in with a band-application system. Although band application should reduce ammonia losses through volatilisation, it is possible that a widely spaced band of fertilizer may not get the N close enough to some plants for them to use this nutrient.

The potential for N losses through ammonia volatilisation can be reduced by using a urease inhibitor to slow or delay hydrolysis. Slowing or delaying this process will allow more time for rainfall to move the urea into the soil. One of the most effective inhibitors currently available is nBTPT, marketed under the trade name Agrotain· The effectiveness of Agrotain depends upon soil and atmospheric conditions at and after the application of a urea-containing fertilizer. A study was conducted in 1996 and 1997 at the Wiregrass Substation in Headland on wheat and ryegrass to evaluate the effectiveness of the urease inhibitor Agrotain when applied with prilled urea and to compare band with broadcast applications of UAN. Agrotain was incorporated onto prilled urea at the labeled rate. Prilled urea, prilled urea plus Agrotain, and prilled ammonium nitrate were broadcast-applied in early March at the rates of 30, 60, and 90 pounds N per acre.

In addition, UAN was applied at the same rates using broadcast and band (10-inch spacing between dribble points) methods. Rain-free periods after application were six days and five days in 1996 and 1997, respectively. The March application is probably too late for maximising yields, but weather patterns were not favourable for ammonia volatilisation until this time. Flag-leaves were sampled from wheat at early heading, and grain was combined at harvest. Ryegrass was clipped in May for dry matter yield and N content.

Wheat grain yields followed a typical response to N rate. Band applications of UAN resulted in higher leaf-N levels than broadcast applications, which indicated that the band applications were more efficient. However, this efficiency was not reflected in grain yields. Leaf-N levels were higher with urea-plus-Agrotain treatment than urea alone, but again this efficiency was not reflected in yields.

Ryegrass yields also followed a typical response to N application rate. Like wheat, the responses were the same for each source in both years, so data in the figure were averaged over years and N sources. Although tissue N levels were higher when UAN was applied in a band, broadcast applications resulted in higher yields. Again, yield response to N sources was the same across N rates and years, so the data in the table are averaged for each source over all rates and both years. Nitrogen levels in the ryegrass tissue were higher with urea plus Agrotain over those from urea alone, indicating increased efficiency when Agrotain was used. However, higher yields were found when urea alone was used.

Table:1. Wheat and Ryegrass Yields as Affected by N Source Averaged Over Application Rates

Nitrogen Source	*Wheat Yield lbs./acre*	*Ryegrass Yield bu./acre*
UAN broadcast	43	3,148
UAN band	43	2,759
Prilled urea	42	2,933
Prilled urea + Agrotain	42	2,484
Prilled ammonium nitrate	45	2,990

In conclusion, the study did not show any yield advantage of using Agrotain applied with prilled urea. According to the data, broadcast applications of UAN to ryegrass were actually better than applying it in a 10-inch band. However, wheat yields did not vary according to the application method of UAN. Rain-free days after application certainly favored N loss through ammonia volatilisation. Other soil and climatic conditions were obviously not favorable, because under favourable conditions, yields with prilled urea should have been lower than those with ammonium nitrate, especially at the lower rates.

UREA

Urea is an organic compound with the chemical formula $(NH_2)_2CO$.Urea is also known by the International Nonproprietary Name (rINN) carbamide, as established by the World Health Organisation. For example, the medicinal compound *hydroxyurea* (old British Approved Name) is now *hydroxycarbamide*. Other names include carbamide resin, isourea, carbonyl diamide, and carbonyldiamine.

Discovery

It was the first organic compound to be artificially synthesised from inorganic starting materials, in 1828 by Friedrich Wöhler, who prepared it by the reaction of potassium cyanate with ammonium sulfate. Although Wöhler was attempting to prepare ammonium cyanate, by forming urea, he inadvertently discredited vitalism, the theory that the chemicals of living organisms are fundamentally different from inanimate matter, thus starting the discipline of organic chemistry.

It is found in mammalian and amphibian urine as well as in some fish. Birds and reptiles excrete uric acid, comprising a different form of nitrogen metabolism that requires less water.

Structure

Urea is highly soluble in water and is, therefore, an efficient way for the human body to expel excess nitrogen. Due to extensive hydrogen bonding with water (up to six hydrogen bonds may form – two from the oxygen atom and one from each hydrogen) urea is very soluble.

The urea molecule is planar and retains its full molecular point symmetry, due to conjugation of one of each nitrogen's P orbital to the carbonyl double bond. Each carbonyl oxygen atom accepts four N-H-O hydrogen bonds, a very unusual feature for such a bond type. This dense (and energetically favourable) hydrogen bond network is probably established at the cost of efficient molecular packing: The structure is quite open, the ribbons forming tunnels with square cross-section.

Endogenous Production

The individual atoms that make up a urea molecule come from carbon dioxide, water, aspartate, and ammonia in a metabolic pathway known as the urea cycle, an anabolic process. Organisms synthesise urea from ammonia because ammonia (a common metabolic waste product) raises pH in cells to toxic levels. Therefore, urea synthesis is necessary even though it costs energy to produce. Urea is neither acidic nor basic, so it is a perfect vehicle for getting rid of nitrogen waste. Urea production occurs in the liver and is regulated by N-acetylglutamate.

In this cycle, amino groups donated by ammonia and L-aspartate are converted to urea, while L-ornithine, citrulline, L-argininosuccinate, and L-arginine act as intermediates.

Function

In Humans

Urea is, in essence, a waste product. However, it also plays a very important role in that it helps set up the countercurrent system in the nephrons. The countercurrent system in the nephrons allows for reabsorption of water and critical ions. Urea is reabsorbed in the inner medullary collecting ducts of the nephrons, thus raising the osmolarity in the medullary interstitium surrounding the thin ascending limb of the Loop of Henle. The greater the osmolarity of the medullary interstitium surrounding the thin ascending Loop of Henle, the more water will be reabsorbed out of the renal tubule back into the interstitium (and thus back into the body). Some of the urea from the meduallary interstitium that helped set up the Countercurrent System will also flow back into the tubule, through urea transporter 2, into the thin ascending limb of the Loop of Henle, through the collecting ducts, and eventually out of the body as a component of urine.

It is dissolved in blood (in a concentration of 2.5 - 7.5 mmol/liter) and excreted by the kidney as a component of urine. In addition, a small amount of urea is excreted (along with sodium chloride and water) in sweat.

Regulation

Control of urea by antidiuretic hormone allows the body to create hyperosmotic urine (urine that has more ions in it—is "more concentrated"—than that same person's blood plasma). Preventing the loss of water in this manner is important if the person's body needs to save water in order to maintain a suitable blood pressure or (more likely,) in order to maintain a suitable concentration of sodium ions in the blood plasma.

Non-humans

Most organisms have to deal with the excretion of nitrogen waste originating from protein and amino acid catabolism. In aquatic organisms the most common form of nitrogen waste is ammonia, while land-dwelling organisms convert the toxic ammonia to either urea or uric acid. In general, birds and saurian reptiles excrete uric acid, whereas the remaining species, including mammals, excrete urea. It is noteworthy that tadpoles excrete ammonia, and shift to urea production during metamorphosis.

Despite the generalisation above, the pathway has been documented not only in mammals and amphibians but in many other organisms as well, including birds, invertebrates, insects, plants, yeast, fungi, and even microorganisms.

Hazards

Urea can be irritating to skin and eyes. Too high concentrations in the blood can cause damage to organs of the body. Low concentrations of urea such as in urine are not dangerous.

It has been found that urea can cause algal blooms to produce toxins, and urea in runoff from fertilizers may play a role in the increase of toxic blooms.

Repeated or prolonged contact with urea in fertilizer form on the skin may cause dermatitis. The substance also irritates the eyes, the skin, and the respiratory tract. The substance decomposes on heating above melting point, producing toxic gases, and reacts violently with strong oxidants, nitrites, inorganic chlorides, chlorites and perchlorates, causing fire and explosion hazard.

Synthetic Production

Urea is a nitrogen-containing chemical product that is produced on a scale of some 100,000,000 tons per year worldwide.

For use in industry, urea is produced from synthetic ammonia and carbon dioxide. Urea can be produced as prills, granules, flakes, pellets, crystals, and solutions.

More than 90% of world production is destined for use as a fertilizer. Urea has the highest nitrogen content of all solid nitrogenous fertilizers in common use (46.7%). Therefore, it has the lowest transportation costs per unit of nitrogen nutrient.

Urea is highly soluble in water and is, therefore, also very suitable for use in fertilizer solutions (in combination with ammonium nitrate: UAN), e.g., in "foliar feed" fertilizers.

Solid urea is marketed as prills or granules. The advantage of prills is that, in general, they can be produced more cheaply than granules, which, because of their narrower particle size distribution, have an advantage over prills if applied mechanically to the soil. Properties such as impact strength, crushing strength, and free-flowing behaviour are, in particular, important in product handling, storage, and bulk transportation.

Commercial Production

Urea is commercially produced from two raw materials, ammonia, and carbon dioxide. Large quantities of carbon dioxide are produced during the manufacture of ammonia from coal or from hydrocarbons such as natural gas and petroleum-derived raw materials. This allows direct synthesis of urea from these raw materials.

The production of urea from ammonia and carbon dioxide takes place in an equilibrium reaction, with incomplete conversion of the reactants. The various urea processes are characterised by the conditions under which urea formation takes place and the way in which unconverted reactants are further processed.

Unconverted reactants can be used for the manufacture of other products, for example ammonium nitrate or sulfate, or they can be recycled for complete conversion to urea in a total-recycle process.

Two principal reactions take place in the formation of urea from ammonia and carbon dioxide. The first reaction is exothermic:

$$2\ NH_3 + CO_2 \rightarrow H_2N\text{-}COONH_4 \text{ (ammonium carbamate)}$$

Whereas the second reaction is endothermic:

$$H_2N\text{-}COONH_4 \rightarrow (NH_2)_2CO + H_2O$$

Both reactions combined are exothermic.

The process, developed in 1922, is also called the Bosch-Meiser urea process after its discoverers.

Uses

Agricultural Use

Urea is used as a nitrogen-release fertilizer, as it hydrolyses back to ammonia and carbon dioxide, but its most common impurity, biuret, must be present at less than 2%, as it impairs plant growth. It is also used in many multi-component solid fertilizer formulations. Its action of nitrogen release is due to the conditions favouring the reagent side of the equilibriums, which produce urea.

Urea is usually spread at rates of between 40 and 300 kg/ha, but actual spreading rates will vary according to farm type and region. It is better to make several small to medium applications at intervals to minimise leaching losses and increase efficient use of the N applied, compared with single heavy applications. During summer, urea should be spread just before, or during rain to reduce possible losses from volatilisation (process wherein nitrogen is lost to the atmosphere as ammonia gas). Urea should not be mixed for any length of time with other fertilizers, as problems of physical quality may result.

Because of the high nitrogen concentration in urea, it is very important to achieve an even spread. The application equipment must be correctly calibrated and properly used. Drilling must not occur on contact with or close to seed, due to the risk of germination damage. Urea dissolves in water for application as a spray or through irrigation systems.

In grain and cotton crops, urea is often applied at the time of the last cultivation before planting. It should be applied into or be incorporated into the soil. In high rainfall areas and on sandy soils (where nitrogen can be lost through leaching) and where good in-season rainfall is expected, urea can be side- or top-dressed during the growing season. Top-dressing is also popular on pasture and forage crops. In cultivating sugarcane, urea is side-dressed after planting, and applied to each ratoon crop.

In irrigated crops, urea can be applied dry to the soil, or dissolved and applied through the irrigation water. Urea will dissolve in its own weight in water, but it becomes increasingly difficult to dissolve as the concentration increases. Dissolving urea in water is endothermic, causing the temperature of the solution to fall when urea dissolves.

As a practical guide, when preparing urea solutions for fertigation (injection into irrigation lines), dissolve no more than 30 kg urea per 100 L water.

In foliar sprays, urea concentrations of 0.5% – 2.0% are often used in horticultural crops. As urea sprays may damage crop foliage, specific advice should be sought before use. Low-biuret grades of urea should be used if urea sprays are to be applied regularly or to sensitive horticultural crops.

Storage of Urea Fertilizer

Like most nitrogen products, urea absorbs moisture from the atmosphere. Therefore it should be stored either in closed/sealed bags on pallets, or, if stored in bulk, under cover with a tarpaulin. As with most solid fertilizers, it should also be stored in a cool, dry, well-ventilated area.

Industrial Use

Urea has the ability to form "loose compounds", called clathrates, with many organic compounds. The organic compounds are held in channels formed by interpenetrating helices comprising of hydrogen-bonded urea molecules. This behaviour can be used to separate mixtures, and has been used in the production of aviation fuel and lubricating oils. As the helices are interconnected, all helices in a crystal must have the same "handedness". This is determined when the crystal is nucleated and can thus be forced by seeding. This property has been used to separate racemic mixtures.

Laboratory Use

Urea is a powerful protein denaturant. This property can be exploited to increase the solubility of some proteins. For this application, it is used in concentrations up to 10 M. Urea is used to effectively disrupt the non-covalent bonds in proteins. Urea is an ingredient in the synthesis of urea nitrate. Urea nitrate is also a high explosive very similar to ammonium nitrate, however it may even be more powerful because of its complexity. VOD is 11,000 fps to 15,420 fps.

Medical Use

Urea is used in topical dermatological products to promote rehydration of the skin. If covered by an occlusive dressing, 40% urea preparations may also be used for nonsurgical debridement of nails. This drug is also used as an earwax removal aid.The blood urea nitrogen (BUN) test is a measure of the amount of nitrogen in the blood that comes from urea. It is used as a marker of renal function.

Other Diagnostic Use

Isotopically-labelled urea (carbon-14 - radioactive, or carbon-13 - stable isotope) is used in the urea breath test, which is used to detect the presence of the bacteria *Helicobacter pylori* (*H. pylori*) in the stomach and duodenum of humans. The test detects the characteristic enzyme urease, produced by *H. pylori*, by a reaction that produces ammonia from urea. This increases pH (reduces acidity) of the stomach environment around the bacteria. Similar bacteria species to *H. pylori* can be identified by the same test in animals such as apes, dogs, and cats (including big cats).

EFFECT OF PLANT DENSITY AND UREA FOLIAR APPLICATION ON YIELD AND YIELD COMPONENTS OF CHICKPEA

Chickpea is considered as one of the most important seed legumes worldwide, i.e., Asia and North Africa. In Egypt it ranks third after faba bean and lentil from the stand point of its importance as legume crop.

It is an important source of cheap protein with high energy and nutritive value; chickpea seeds consumed dry by parched or cooked in human diets, in sweet manufactures and recently in baby food blends, El Kramany and Bahr. But detailed research on sowing rate and plant population density (PPD) has not been conducted. Plant population density has great effect on yield and yield components of chickpea, many researchers reported the effect of plant population density on chickpea. Jettner *et al.* stated that increasing yields at high sowing rate can be directly attributed to large plant population. Ali and Singh reported that the effect of population densities 33 plants/m^2 and 40 plant/m^2 do not differ significantly for yield as well as yield attributes. Concluded that there was strong relationship between economic optimum plant density and seed yield potential, in higher yielding situations (>1.5 t/ha) plant densities (>35 plant/m^2) will produce the most profit.

Pointed that plant population density (PPD) produced the highest seed yields ranged from 40 to 50 plant/m^2 at different types of chickpea. Concluded that the total pod production of chickpea crop depend on plants/m^2 more than on pods/plant, Also seed yield potential of chickpea would be increased by increasing plant population to produce more pods per unit area.

Stated that sowing chickpea at 25 seeds/m^2 yielded bigger seeds, which usually fetch higher prices. Frade and Valenciano pointed that there was a reduction in the number of pods/plant (mean, from 56 at 8 plants/m^2 to 14 at 36 plant/m^2) and the 1000 seed weight (mean from 406 g at 8 plant/m^2 to 383 g at 36 plants/m^2) as plant density was increased.

The highest yields of 222g/m^2 were obtained at the highest density 36 plants/m^2. Stated that any restriction to the photosyntheisis due to the depletion of nitrogen in the leaves during the pod-filling period due to poor nutrition uptake from soil and translocation of this element from leaves to the developing seeds may lead to the acceleration of leaf senescence.

The positive effect of supplying legume plants with supplementary nitrogen was found to be with beneficial effect on enhancing growth and increasing seed yield concluded that foliar application of urea at 50% flowering increased yield and seed protein.

An increase in number of pods with more than one seed rather than from increased pod number/plant or increased seed size. Also, the increase in the seed protein content resulted from increased nitrogen availability for seed filling.

The results indicate the potential to increase yields of chickpea by application of foliar nitrogen near flowering. Therefore, this work studies the effect of both plant population density (PPD) and urea foliar application on yield and yield components of chickpea.

Materials and Methods

The experimental soil was analysed according to the method described by Chapman and Pratt. Soil texture was sandy and having the following characteristics: Sand 94%; pH 8.3; organic matter 3 0.85%; CaCo 0.35%; EC 0.07 mmhos/cm^3; total N_3.1 mg N/100 g and 1.7 mg P/100g. Chickpea seeds c.v Giza 531 previously inoculated with the specific strain of *Rhizobium leguminosarum* were sown on the second week of November in both seasons in plots 10.5 m^2 (5 ridges, each 3 m in length and 0.70 m in width) in hills 2 5 20 cm apart.

Calcium superphosphate 15.5% P O at 2 the rate of 32 kg P O5/fed. and N starter dose at the rate of 5Kg N/fed in the form of ammonium nitrate 2 and 10 Kg K 0/fed. as potassium sulphate were applied before ridging. Three weeks after sowing plants were thinned to two plants/hill, then 15 kg N/fed. as ammonium nitrate 33% N were applied. The experimental design was complete randomized block design in three replicates. Treatments were combination between 2 plant densities and 4 dates of urea foliar application. The experimental treatments can be described as follows:

A - Plant density.

A -1- Low density 26 plants/m^2

A -2- High density 50 plants/m^2

Low density sown seeds in one side of ridge and high density sown seeds in both sides of ridge. Number of plants/m^2 was determined by counting the plants in 1 m^2 from each plot after complete emergence.

B - Urea foliar application.

B -1- control (foliar application with tap water).

B -2- 1% urea foliar application at flowering stage.

B -3- 1% urea foliar application at pod set stage.

B -4- 1% urea foliar application at seed filling stage.

Addition of 1.84 Kg N/fed. (1% urea) the volume of the spray solution was 400 L/fed. Teepol as a surfactant was added to the spray solution at a rate of 1 ml/L. control plants were sprayed by 400L water only. At harvest 120 days after sowing (DAS) two central ridges were harvested and sub samples of 10 plants, were taken randomly to estimate the yield attributes as follows:

- Plant height (cm).
- Number of pods/plant.
- Number of seeds/plant 4-1000 seeds weight.
- Total dry matter (g).

The whole plot was harvested to determine seed and biological yields/ fed. then calculated harvest index by divided seed yield/biological yield. Total N-content in seeds determined and protein % was calculated by multiplying N-content by 6.25 according to Chapman and Pratt. All data were statistically analysed and combined analysis was conducted for the data of two seasons according to Snedecor and Cochran. The least significant difference (LSD) was used to compare between means.

Effect of Plant Density

Low density (26 plant/m^2) and high density (50 plant/m^2) for yield, yield attributes, N and protein per cent in chickpea seeds. It is clear that the differences were significantly for all studied attributes except plant height, number of branches/plant, no. of seeds/plant, N% and protein per cent in seeds.

The high density 50 plant/m^2 produced taller plants, bigger branches, pods, seeds number than low density 26 plants/m^2, also, seed yield/plant per, feddan and larger biological yield/fed., higher harvest index.

Chickpea seeds sown by high density contain higher N, protein per cent than low density, this results were in accordance with those obtained by Jettner who reported that increasing yield at high density can be directly attributed to large population, also, they found strong relationship between seed yield and plant population densities. Concluded that the total pod production of chickpea crop depend on plants/m^2 more than on pods/plant, also, seed yield potential of chickpea would be increased by increasing plant population to produce more pods per unit area.

Effect of Urea Foliar Application

The differences between time of urea foliar application to chickpea plants at flowering, pod set and pod filling compared to control (untreated). Data show that there were significant differences between treatments for yield and all yield attributes. Treatments of 1% urea foliar application at pod filling gave the tallest plants, having the highest number of branches, pods, seeds/plant, the heaviest 100 seed in gram.

The same treatment came in the first order due to plant total dry weight. Arrangement of urea foliar application treatments in descending order was pod filling - pod set - flowering - control for seed yield/fed., biological yield/fed., N% in seeds and protein% in seeds. The best harvest index recorded by 1 per cent urea foliar application at pod filling stage followed by pod set, control and at flowering stage.

The same result was found by Behairy who indicated that any striction to the photosynthesis due to the depletion of nitrogen in the leaves during the pod filling period. Increasing seed and biological yield as a result of urea foliar application was reported by Zeidan; El Kramany and Mirvat and Palta *et al.* who concluded that the increase in seed protein content resulted from increased nitrogen availability for seed filling.

Interaction of Plant Density X Urea Foliar Application

Results indicated that the differences were significant in 100 seed weight; seed yield either per plant or per feddan; biological yield/fed; harvest index; N per cent in seeds and protein per cent in seeds. Interaction of high density 50 plants/m^2 × 1% urea foliar application at pod filling recorded the tallest plants which produced the highest number of branches, pods, seeds/plant, the heaviest 100 seed, the highest seed yield either per plant or per feddan; biological yield/fed.; harvest index; N% in seeds and protein per cent in seeds.

Chapter 10

Fertilizers as Water Pollutants

EUTROPHICATION OF SURFACE WATERS

"Eutrophication" is the enrichment of surface waters with plant nutrients. While eutrophication occurs naturally, it is normally associated with anthropogenic sources of nutrients. The "trophic status" of lakes is the central concept in lake management. It describes the relationship between nutrient status of a lake and the growth of organic matter in the lake. Eutrophication is the process of change from one trophic state to a higher trophic state by the addition of nutrient.

Agriculture is a major factor in eutrophication of surface waters. The most complete global study of eutrophication was the Organisation for Economic Cooperation and Development (OECD) Cooperative Programme on Eutrophication carried out in the 1970s in eighteen countries.

The sequence of trophic state, from oligotrophic (nutrient poor) to hypertrophic (= hypereutrophic [nutrient rich]). Although both nitrogen and phosphorus contribute to eutrophication, classification of trophic status usually focuses on that nutrient which is limiting. In the majority of cases, phosphorus is the limiting nutrient. While the effects of eutrophication such as algal blooms are readily visible, the process of eutrophication is complex and its measurement difficult.

	Phosphorus	*Chlorophyll*
Ultra-oligotrophic	10%	6%
Oligotrophic	63%	49%
Mesotrophic	26%	42%
Eutrophic	1%	3%
Hypertrophic	0%	0%

Because of the complex interaction amongst the many variables that play a part in eutrophication, Janus and Vollenweider concluded that it is impossible to develop strict boundaries between trophic classes. They calculated, for example, the probability (as per cent) of classifying a lake with total phosphorus and chlorophyll-a concentrations of 10 and 2.5 mg/m^3 respectively, as:

The symptoms and impacts of eutrophication are:

- Increase in production and biomass of phytoplankton, attached algae and macrophytes.
- Shift in habitat characteristics due to change in assemblage of aquatic plants.
- Replacement of desirable fish (e.g., salmonids in western countries) by less desirable species.
- Production of toxins by certain algae.
- Increasing operating expenses of public water supplies, including taste and odour problems, especially during periods of algal blooms.
- Deoxygenation of water, especially after collapse of algal blooms, usually resulting in fish kills.
- Infilling and clogging of irrigation canals with aquatic weeds (water hyacinth is a problem of introduction, not necessarily of eutrophication).
- Loss of recreational use of water due to slime, weed infestation and noxious odour from decaying algae.
- Impediments to navigation due to dense weed growth.
- Economic loss due to change in fish species, fish kills, etc.

ROLE OF AGRICULTURE IN EUTROPHICATION

In their summary of water quality impacts of fertilizers, FAO/ECE cited the following problems:

- Fertilization of surface waters (eutrophication) results in, for example, explosive growth of algae which causes disruptive changes to the biological equilibrium. This is true both for inland waters (ditches, river, lakes) and coastal waters.
- Groundwater is being polluted mainly by nitrates. In all countries groundwater is an important source of drinking water. In several areas the groundwater is polluted to an extent that it is no longer fit to be used as drinking water according to present standards.

While these problems were primarily attributed to mineral fertilizers by FAO/ECE, in some areas the problem is particularly associated with extensive and intensive application of organic fertilizers.

The precise role of agriculture in eutrophication of surface water and contamination of groundwater is difficult to quantify. Where it is warranted, the use of environmental isotopes can aid in the diagnosis of pollutant pathways to and within groundwater. RIVM, citing Isermann calculated that European agriculture is responsible for 60% of the total riverine flux of nitrogen to the North Sea and 25% of the total phosphorus loading. Agriculture also makes a substantial contribution to the total atmospheric nitrogen loading to the North and the Baltic Seas. This amounts to 65% and 55% respectively.

Czechoslovakia reported that agriculture contributes 48% of the pollution of surface water; Norway and Finland reported locally significant eutrophication of surface waters arising from agriculture; high levels of usage of N and P are considered to be responsible for proliferation of algae in the Adriatic; similar observations are made in Danish coastal waters; substantial contamination of groundwater by nitrate in the Netherlands was also reported. Appelgren reported that 50% of shallow groundwater wells that supply more than one million rural residents in Lithuania are unfit for human consumption because of a wide range of pollutants which include pesticides and nitrogen species. In the 1960s Lake Erie was declared "dead" by the press due to the high levels of nutrients accompanied by excessive growth of algae, fish kills and anaerobic bottom sediments.

Although the ECE regarded livestock wastes as a point source and excluded it from calculations of the contribution of agriculture to eutrophication in Europe, their statistics indicated that livestock wastes accounted "on average" for 30% of the total phosphorus load to European inland waters, with the rest of agriculture accounting for a further 17%. The situation for nitrogen, as for phosphorus, was quite variable from country to country. Danish statistics indicated that manure contributes at least 50% of the leaching of inorganic N. Nitrogen from agricultural non-point sources in the Netherlands amounted to 71% of the total N load generated from within the Netherlands.

A study by Ryding in Sweden demonstrated how lakes which were unaffected by industrial or municipal point sources, underwent long-term change in nutrient status as a result of agricultural activities in the watershed. Over the period 1973-1981 the nutrient status of Lake Oren increased from 780 to 1000 mg/m3 for Total-N and from 10 to 45 mg/m^3 for Total-P. Lake transparency declined from 6.2 to 2.6 m and suffered periodic (heavy) algal blooms.

Heavily fertilized crops such as maize tend to have large losses relative to non-intensive uses such as pasture. Agricultural uses associated with poor land management practices that lead to erosion also produce significant nutrient losses. Wastes, manures and sludges, through biological concentration processes, can supply soils with 100 times more hazardous products than fertilizers for the equivalent plant nutrient content. This is considered a major environmental (and water quality) problem in periurban areas of many developing countries. Numerous authors report that a high degree of variability at individual sites is expected as a consequence of changes in hydrological regime from year to year. The implication is that estimation techniques using "typical" values of nutrient yield can expect to have a high degree of uncertainty and could be very much in error if estimated from data collected over a single year.

There is a danger, however, in assuming that all waters have natural levels that are low in nutrients. In some areas, such as lakes located in areas of rich agricultural soils, waters have historically been highly enriched by nutrients associated with natural erosion of fertile soils. In the prairie lakes of Canada, for example, early settlers reported that the lakes were green with algae.

In other parts of the world, as in Asia, ancient civilisations so profoundly impacted water quality that there are no longer "natural" levels of nutrients. In such situations the existence of eutrophication, while undeniable, must be measured against arbitrary standards that reflect water quality criteria established on the basis of societal needs for beneficial use of the water.

ORGANIC FERTILIZERS

The importance and, in some cases, the major problems associated with organic fertilizers, deserve special mention. Manure produced by cattle, pigs and poultry are used as organic fertilizer the world over. To this is added human excreta, especially in some Asian countries where animal and human excreta are traditionally used in fish culture as well as on soils. However, intensive livestock production has produced major problems of environmental degradation, a phenomenon which has been the subject of European and North American legislation and control. The problem is particularly acute in areas of intensive livestock production, such as in the Eastern and Southern parts of the Netherlands where the production of manure greatly exceeds the capacity of the land to assimilate these wastes.

In addition to problems associated with excessive application of manure on the land, is the problem of direct runoff from intensive cattle, pig and poultry farms. Although this is controlled in many Western countries, it constitutes a serious problem for water quality in much of the rest of the world.

For example, Appelgren reports that discharge of pig wastes from intensive pig raising in Lithuania is a major source of surface water pollution in that country. The FAO/ECE reports similar problems in the Po River of Italy. The Canadian Department of Agriculture calculated in 1978, on the basis of detailed study of several feedlot operations, that cattle feedlots and manure storage facilities contributed 0.5-13% of the total loading of total phosphorus at that time to the Canadian portion of the Lower (agricultural portion) Great Lakes.

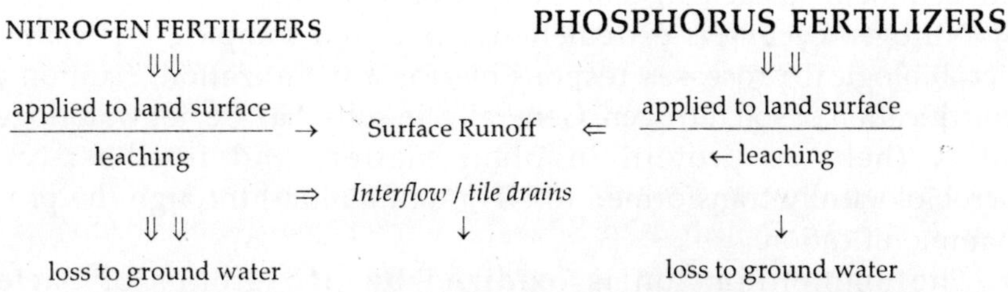

Fig:1. Schematic Diagram of Nitrogen and Phosphorus Losses
(Arrows are proportional to loss)

To the typical pathways of degradation, that of surface runoff and infiltration into the groundwater, is added the volatilization of ammonia which adds to acidification of land and water. In a review of environmental impacts caused by animal husbandry in Europe, the FAO/ECE reported the following major categories of impacts:

- Fertilization of surface waters, both as a result of direct discharges of manure and as a consequence of nitrate, phosphate and potassium being leached from the soil.
- Contamination of the groundwater as a result of leaching, especially by nitrate. Phosphates are less readily leached out, but in areas where the soil is saturated with phosphate this substance is found in the groundwater more and more often.
- Surface waters and the groundwater are being contaminated by heavy metals. High concentrations of these substances pose a threat to the health of man and animals. To a certain extent these heavy metals accumulate in the soil, from which they are taken up by crops. For example, pig manure contains significant quantities of copper.
- Acidification as a result of ammonia emission (volatilization) from livestock accommodation, manure storage facilities and manure being spread on the land. Ammonia constitutes a major contribution to the acidification of the environment, especially in areas with considerable intensive livestock arming.

ENVIRONMENTAL CHEMISTRY

The key hydrological processes that link rainfall, runoff and leaching and which give rise to erosion and transport of chemically enriched soil particles, are important components of the environmental chemistry, transport and fate of fertilizer products. The environmental dynamics of nitrogen and phosphorus are well known although the detailed transformations of nitrogen that occur in soil and water are difficult to study and document. Nitrogen is comprised of the forms: soluble organic N, NH_4-N (ammonium), NO_3-N (nitrate), NO_2-N (nitrite) and N associated with sediment as exchangeable NH_4-N or organic-N.

Nitrogen cycling is extremely dynamic and complex, especially the microbiological processes responsible for mineralization, fixation and denitrification of soil nitrogen. Generally, in soils that are not waterlogged, soil N (held as protein in plant matter) and fertiliser-N are microbiologically transformed to NH_4 (ammonium) through the process of ammonification.

The ammonium ion is oxidized by two groups of bacteria (Nitrosomonas and Nitrobacter) to NO_3 with an unstable intermediate NO_2 product in a process called nitrification. Urea is readily hydrolysed to ammonium. Denitrification occurs under anoxic conditions such as wetlands where NO_3 is reduced to various gaseous forms. The N cycle is largely controlled by bacteria, hence the rate of N cycling is dependent upon factors such as soil moisture, temperature, pH, etc. NO_3 is the end-product of aerobic N decomposition and is always dissolved and mobile.

From a water quality perspective, the ammonium ion (NH_4) can be adsorbed to clay particles and moved with soil during erosion. More importantly, however, NH_4 and NO_3 are soluble and are mobilised through the soil profile to groundwater during periods of rain by the process of leaching. NO_3 is also observed in surface runoff during rainfall events.

Prevention of nitrogen pollution of surface and groundwater depends very much on the ability to maintain NO_3 in soil only up to the level that can be taken up by the crop and to reduce the amount of NO_3 held in the soil after harvesting.

In contrast, the behaviour of phosphorus is quite simple. Phosphorus can exist in a variety of forms: as mineral (generally apatite) phosphorus (AP); non-apatite inorganic-P (NAIP); organic-P (OP - bound up with carbon and oxygen in plant matter); and as dissolved soluble reactive ortho-P (SRP). The phosphorus species AP, NAIP and OP are associated with the particulate phase. In studies of phosphorus movement from agricultural lands the largest amount is sorbed onto clay materials and transported as erosion products.

SRP is readily available to aquatic plants to the point where measured SRP in surface water may only represent a residual amount after most of the SRP has been taken up by plant life. Consequently, in aquatic studies, the focus is often on the sediment-associated forms of P as these tend to dominate total phosphorus flux.

The NAIP fraction is considered to be available to plant roots and is rapidly solubilised under conditions of anoxia in the bottom of lakes and reservoirs. It is for this reason that lake sediments can represent a very large internal (autotrophic) load of phosphorus which is recycled into the water column during periods of bottom anoxia. This load can be so large that, without attention to lake sediment remediation, phosphorus management programmes in tributaries can be quite meaningless.

The Point Versus Non-point Source Dilemma

The dilemma in many countries is to ascertain the role of agriculture relative to the impacts of (often untreated) municipal sewage. In a large number of countries the database required to make this distinction is lacking and frustrates the development of a rational pollution abatement programme and inhibits cost-effective investment in control measures. In developing countries it makes sense that the focus should initially be on point source control; however, it has been the experience in the developed countries that point source control for nutrients has not had the desired level of environmental benefit until agricultural control measures were seriously addressed. It is significant that the trend of fertilizer usage worldwide has been one of huge increases in the past 40 years suggesting that, in the absence of major changes in land use to control fertilizer runoff in large parts of the world, one may expect that agriculture will be responsible for an ever-increasing contribution to surface water pollution.

The observations reported by Quirós for the La Plata basin are indicative of the difficulty in segregating the effects of agriculture from other sources. In the Great Lakes of North America some $10 million dollars were spent between 1970 and 1980 to quantify the relative impacts of point versus non-point sources.

That exercise proved enormously successful and specific policies were adopted for nutrient control in each lake basin that reflected the relative contributions from each type of source.

MANAGEMENT OF WATER QUALITY IMPACTS FROM FERTILIZERS

The investigation of eutrophication of surface water by agriculture must adopt a pragmatic management perspective. Of value to agriculture is the perspective adopted by the OECD study of eutrophication. That study focused on the following aspects:

- The qualitative assessment of the trophic state of bodies of water in terms of a few easily measured parameters.
- The dependence of this state on nutritional conditions and nutrient load.
- Translation of these results to the needs of eutrophication control for management.

The progression of these aspects is interesting in that the focus is on easily measurable state of the water body, followed by a determination of the extent to which the state is a product of nutrient loads, then the degree to which loads may be manipulated to achieve a desired trophic state that is determined by water use. Prediction of water quality impacts of fertilizers and related land management practices is an essential element of site-specific control options and for the development of generic approaches for fertilizer control. Prediction tools are essentially in the form of models.

Mineral Fertilizers

The response to the need to control leaching and runoff of nutrients and contamination of soils and water by heavy metals has been variable in Europe. Control measures are part of the larger issue of mineral and organic fertilizer usage. FAO/ECE summarized the types of voluntary and mandated controls in Europe that apply to mineral fertilizers as:

- Taxes on fertilizer.
- Requirement for fertilizer plans.
- Preventing the leaching of nutrients after the growing season by increasing the area under autumn/winter green cover and by sowing crops with elevated nitrogen.
- Promoting and subsidising better application methods, developing new, environmentally sound fertilizers and promoting soil testing.
- Severely limiting the use of fertilizers in e.g., water extraction areas and nature protection areas.

In any location where intensive agriculture and/or livestock farming produces serious risks of nitrogen pollution, Ignazi recommended the following essential steps that are taken at the farm level:

Rational Nitrogen Application

To avoid over-fertilization, the rate of nitrogen fertilizer to be applied needs to be calculated on the basis of the "crop nitrogen balance". This takes into account plant needs and amount of N in soil.

Vegetation Cover

As far as possible, keep the soil covered with vegetation. This inhibits build-up of soluble nitrogen by absorbing mineralised nitrogen and preventing leaching during periods of rain.

Manage the Period between Crops

Organic debris produced by harvesting is easily mineralized into leachable N. Steps to reduce leachable N includes planting of "green manure" crops and delaying ploughing of straw, roots and leaves into the soil.

Rational Irrigation

Poor irrigation has one of the worst impacts on water quality, whereas precision irrigation is one of the least polluting practices as well as reducing net cost of supplied water.

Optimise other Cultivation Techniques

Highest yields with minimum water quality impacts require optimisation of practices such as weed, pest and disease control, liming, balanced mineral fertilizers including trace elements, etc.

Agricultural Planning

Implement erosion control techniques that complement topographic and soil conditions.

Organic Fertilizers

Voluntary and legislated control measures in Europe are intended to have the following benefits:

- Reduce the leaching of nutrients.
- Reducing emissions of ammonia.
- Reducing contamination by heavy metals.

The nature of these measures varies by country; however, FAO/ECE have summarised the types of voluntary and mandated control as:

- Maximum numbers of animals per hectare based on amount of manure that can be safely applied per hectare of land.
- Maximum quantities of manure that can be applied on the land is fixed, based on the N and P content of the manure.
- Holdings wishing to keep more than a given number of animals must obtain a license.
- The periods during which it is allowed to apply manure to the land have been limited and it is obligatory to work it into the ground immediately afterwards.
- Establishment of regulations on minimum capacity for manure storage facilities.
- Establish fertilizer plans.
- Levies (taxes) on surplus manure.
- Areas under autumn/winter green cover were extended and green fallowing is being promoted.
- Maximum amounts established for spreading of sewage sludge on land based on heavy metal content.
- Change in composition of feed to reduce amount of nutrients and heavy metals.
- Research and implementation of means of reducing ammonia loss.

Sludge Management

Sludge is mentioned here only insofar as the spreading of sludge from municipal wastewater treatment facilities on agricultural land is one method used to get rid of municipal sludge in a way that is perceived to be beneficial. The alternatives are incineration and land fill. FAO/ECE include sludge within the category of organic fertilizers but note that sludge often contains unacceptable levels of heavy metals. Pollution of water by sludge runoff is otherwise the same as for manure noted above.

ECONOMICS OF CONTROL OF FERTILIZER RUNOFF

Nutrient loss is closely associated with rainfall-runoff events. For phosphorus, which tends to be associated with the solid phase (sediment), runoff losses are directly linked to erosion. Therefore the economics of nutrient control tend to be closely tied to the costs of controlling runoff and erosion. Therefore, this will be treated briefly here. In particular, it is useful to examine the economic cost of nutrient runoff which must be replaced by fertilizers if the land is to remain productive.

The link between erosion, increasing fertilizer application and loss of soil productivity is very direct in many countries. In the Brazilian state of Paraná where agriculture is the base of the state economy, Paraná produces 22% of the national grain production on only 2.4% of the Brazilian territory.

Agricultural expansion in Paraná occurred mainly in the period 1950-1970 and was "characterised by short-term agricultural systems leading to continuous and progressive environmental degradation as a result of economic policies and a totally inappropriate land parcelling and marketing system...". Erosion has led to extensive loss of top soil, large-scale gullying and silting of ditches and rivers. The use of fertilizers has risen as a consequence, up 575% over the period 1970-1986 and without any gain in crop yields. Loss of N-P-K from an average erosion of 20 t/ha/yr represents an annual economic loss of US $242 million in nutrients.

Analysis by Elwell and Stocking of nutrient loss arising from erosion in Zimbabwe shows similar significant economic losses in African situations. Stocking, applying data collected in the 1960s by Hudson to the soil use map of Zimbabwe, calculated an annual loss of 10 million tons of nitrogen and 5 million tons of phosphorus annually as a consequence of erosion.

Roose also cites losses of 98 kg/ha/yr of nitrogen, 29 kg/ha/yr of phosphorus, 39 kg/ha/yr of lime and 39 kg/ha/yr of magnesium from soils of lower Côte d'Ivoire as a result of erosion. This loss is so severe that compensation requires 7 tons of fresh manure annually, plus 470 kg of ammonium sulphate, 160 kg of superphosphate, 200 kg dolomite and 60 kg of potassium chloride per hectare per year.

Roose notes that it is not surprising that the soil is exhausted after only two years of traditional agriculture. Furthermore, these losses do not take into account additional losses due to harvesting and runoff. Roose summarises by stating that action against soil erosion is essential in order to manage what he describes as a "terrible" chemical imbalance in soils caused by soil erosion.

Economic losses tend to be higher in tropical countries where soils, rainfall and agricultural practices are more conducive to erosion and reported rates of erosion are much above average. The World Bank reported that extrapolation from test-plots of impacts of soil loss on agricultural productivity, indicates some 0.5-1.5% loss of GDP annually for countries such as Costa Rica, Malawi, Mali and Mexico. These losses do not include offsite costs such as reservoir infilling, river sedimentation, damage to irrigation systems, etc.

Soil fertility is a complex issue and nutrient loss is not necessarily nor always a consequence of erosion. Erosion and soil loss is the end member of a variety of physical, vegetative and nutrient factors that lead to soil degradation. Global patterns of fertilizer application indicate however that rapidly rising levels of fertilizer utilisation are required merely to maintain soil productivity from a variety of types of loss, including losses due to erosion and, more generally, to soil degradation.

AQUACULTURE

Aquaculture is a special case of agricultural pollution. There are two main forms: land-based and water-based systems. Effluent controls are possible on land-based systems, however water-based systems present particular problems. Aquaculture is rapidly expanding in most parts of the developed and developing world, both in freshwater and marine environments. In contrast, coastal fisheries in most countries are declining.

The environmental impact is primarily a function of feed composition and feed conversion (faecal wastes), plus assorted chemicals used as biocides, disinfectants, medicines, etc. Wastage of feed (feed not taken up by the fish) is estimated to be 20% in European aquaculture. Waste feed and faecal production both add substantial nutrient loadings to aquatic systems.

Additional environmental problems include risk of disease and disease transfer to wild fish, introduction of exotic species, impacts on benthic communities and on the eutrophication of water, interbreeding of escaped cultured fish with wild fish with consequent genetic change in the wild population. Traditional integrated aquaculture systems, as in China, where sewage-fish culture is practised, can be a stabilising influence in the entire ecosystem. This is recommended, especially in developing countries where water and resources are scarce or expensive.

PROBLEMS OF RESTORATION OF EUTROPHIC LAKES

Eutrophic and hypertrophic lakes tend to be shallow and suffer from high rates of nutrient loadings from point and non-point sources. In areas of rich soils such as the Canadian prairies, lake bottom sediments are comprised of nutrient-enriched soil particles eroded from surrounding soils. The association of phosphorus with sediment is a serious problem in the restoration of shallow, enriched lakes. P-enriched particles settle to the bottom of the lake and form a large pool of nutrient in the bottom sediments that is readily available to rooted plants and which is released from bottom sediments under conditions of anoxia into the overlying water column and which is quickly utilised by algae.

This phosphorus pool, known as the "internal load" of phosphorus, can greatly offset any measures taken by river basin managers to control lake eutrophication by control of external phosphorus sources from agriculture and from point sources. Historically, dredging of bottom sediments was considered the only means of remediating nutrient-rich lake sediments, however, modern technology now provides alternative and more cost-effective methods of controlling internal loads of phosphorus by oxygenation and by chemically treating sediments *in situ* to immobilise the phosphorus. Nevertheless, lake restoration is expensive and must be part of a comprehensive river basin management programme.

PESTICIDES AS WATER POLLUTANTS

The term "pesticide" is a composite term that includes all chemicals that are used to kill or control pests. In agriculture, this includes herbicides (weeds), insecticides (insects), fungicides (fungi), nematocides (nematodes) and rodenticides (vertebrate poisons).

A fundamental contributor to the Green Revolution has been the development and application of pesticides for the control of a wide variety of insectivorous and herbaceous pests that would otherwise diminish the quantity and quality of food produce. The use of pesticides coincides with the "chemical age" which has transformed society since the 1950s. In areas where intensive monoculture is practised, pesticides were used as a standard method for pest control. Unfortunately, with the benefits of chemistry have also come disbenefits, some so serious that they now threaten the long-term survival of major ecosystems by disruption of predator-prey relationships and loss of biodiversity. Also, pesticides can have significant human health consequences. While agricultural use of chemicals is restricted to a limited number of compounds, agriculture is one of the few activities where chemicals are intentionally released into the environment because they kill things.

Agricultural use of pesticides is a subset of the larger spectrum of industrial chemicals used in modern society. The American Chemical Society database indicates that there were some 13 million chemicals identified in 1993 with some 500 000 new compounds being added annually. In the Great Lakes of North America, for example, the International Joint Commission has estimated that there are more than 200 chemicals of concern in water and sediments of the Great Lakes ecosystem. Because the environmental burden of toxic chemicals includes both agriculture and non-agricultural compounds, it is difficult to separate the ecological and human health effects of pesticides from those of industrial compounds that are intentionally or accidentally released into the environment. However, there is overwhelming evidence that agricultural use of pesticides has a major impact on water quality and leads to serious environmental consequences.

Irrigated agriculture, especially in tropical and subtropical environments, usually requires modification of the hydrological regime which, in turn, creates habitat that is conducive to breeding of insects such as mosquitoes which are responsible for a variety of vector-borne diseases. In addition to pesticides used in the normal course of irrigated agriculture, control of vector-borne diseases may require additional application of insecticides such as DDT which have serious and widespread ecological consequences. In order to address this problem, environmental management methods to control breeding of disease vectors are being developed and tested in many irrigation projects.

Historical Development of Pesticides

The history of pesticide development and use is the key to understanding how and why pesticides have been an environmental threat to aquatic systems and why this threat is diminishing in developed countries and remains a problem in many developing countries. Stephenson and Solomon outlined the chronology.

North-South Dilemma over Pesticide Economics

As noted above, the general progression of pesticide development has moved from highly toxic, persistent and bioaccumulating pesticides such as DDT, to pesticides that degrade rapidly in the environment and are less toxic to non-target organisms. The developed countries have banned many of the older pesticides due to potential toxic effects to man and/or their impacts on ecosystems, in favour of more modern pesticide formulations.

In the developing countries, some of the older pesticides remain the cheapest to produce and, for some purposes, remain highly effective as, for example, the use of DDT for malaria control. Developing countries maintain that they cannot afford, for reasons of cost and/or efficacy, to ban certain older pesticides.

The dilemma of cost/efficacy versus ecological impacts, including long range impacts via atmospheric transport and access to modern pesticide formulations at low cost remains a contentious global issue. In addition to ecological impacts in countries of application, pesticides that have been long banned in developed countries (such as DDT, toxaphene, etc.), are consistently found in remote areas such as the high arctic.

Chemicals that are applied in tropical and subtropical countries are transported over long distances by global circulation. The global situation has deteriorated to the point where many countries are calling for a global convention on "POPs" (Persistent Organic Pollutants) which are mainly chlorinated compounds that exhibit high levels of toxicity, are persistent and bioaccumulate. The list is not yet fixed; however, "candidate" substances include several pesticides that are used extensively in developing countries.

Fate and Effects of Pesticides

Human Health Effects of Pesticides

Perhaps the largest regional example of pesticide contamination and human health is that of the Aral Sea region. UNEP linked the effects of pesticides to "the level of oncological (cancer), pulmonary and haematological morbidity, as well as on inborn deformities... and immune system deficiencies".

Human health effects are caused by:

- *Skin contact:* Handling of pesticide products.
- *Inhalation:* Breathing of dust or spray.
- *Ingestion:* Pesticides consumed as a contaminant on/in food or in water.

Farm workers have special risks associated with inhalation and skin contact during preparation and application of pesticides to crops. However, for the majority of the population, a principal vector is through ingestion of food that is contaminated by pesticides. Degradation of water quality by pesticide runoff has two principal human health impacts. The first is the consumption of fish and shellfish that are contaminated by pesticides; this can be a particular problem for subsistence fish economies that lie downstream of major agricultural areas.

The second is the direct consumption of pesticide-contaminated water. WHO has established drinking water guidelines for 33 pesticides. Many health and environmental protection agencies have established "acceptable daily intake" (ADI) values which indicate the maximum allowable daily ingestion over a person's lifetime without appreciable risk to the individual. Tetrachlorohydroquinone, a toxic metabolite of the biocide pentachlorophenol, was found to produce "significant and dose-dependent DNA damage".

Ecological Effects of Pesticides

Pesticides are included in a broad range of organic micro-pollutants that have ecological impacts. Different categories of pesticides have different types of effects on living organisms, therefore generalisation is difficult. Although terrestrial impacts by pesticides do occur, the principal pathway that causes ecological impacts is that of water contaminated by pesticide runoff. The two principal mechanisms are bioconcentration and biomagnification.

Bioconcentration

This is the movement of a chemical from the surrounding medium into an organism. The primary "sink" for some pesticides is fatty tissue ("lipids"). Some pesticides, such as DDT, are "lipophilic", meaning that they are soluble in and accumulate in, fatty tissue such as edible fish tissue and human fatty tissue. Other pesticides such as glyphosate are metabolised and excreted.

Biomagnification

This term describes the increasing concentration of a chemical as food energy is transformed within the food chain. As smaller organisms are eaten by larger organisms, the concentration of pesticides and other chemicals are increasingly magnified in tissue and other organs. Very high concentrations can be observed in top predators, including man. The ecological effects of pesticides (and other organic contaminants) are varied and are often inter-related.

Effects at the organism or ecological level are usually considered to be an early warning indicator of potential human health impacts. The major types of effects are listed below and will vary depending on the organism under investigation and the type of pesticide.

Different pesticides have markedly different effects on aquatic life which makes generalisation very difficult. The important point is that many of these effects are chronic (not lethal), are often not noticed by casual observers, yet have consequences for the entire food chain.

- Death of the organism.
- Cancers, tumours and lesions on fish and animals.
- Reproductive inhibition or failure.
- Suppression of immune system.
- Disruption of endocrine (hormonal) system.
- Cellular and DNA damage.

- Teratogenic effects (physical deformities such as hooked beaks on birds).
- Poor fish health marked by low red to white blood cell ratio, excessive slime on fish scales and gills, etc.
- Intergenerational effects (effects are not apparent until subsequent generations of the organism).
- Other physiological effects such as egg shell thinning.

These effects are not necessarily caused solely by exposure to pesticides or other organic contaminants, but may be associated with a combination of environmental stresses such as eutrophication and pathogens. These associated stresses need not be large to have a synergistic effect with organic micro pollutants.

Ecological effects of pesticides extend beyond individual organisms and can extend to ecosystems. Swedish work indicates that application of pesticides is thought to be one of the most significant factors affecting biodiversity. Jonsson report that the continued decline of the Swedish partridge population is linked to changes in land use and the use of chemical weed control.

Chemical weed control has the effect of reducing habitat, decreasing the number of weed species and of shifting the balance of species in the plant community. Swedish studies also show the impact of pesticides on soil fertility, including inhibition of nitrification with concomitant reduced uptake of nitrogen by plants. These studies also suggest that pesticides adversely affect soil micro-organisms which are responsible for microbial degradation of plant matter (and of some pesticides) and for soil structure.

Natural Factors that Degrade Pesticides

In addition to chemical and photochemical reactions, there are two principal biological mechanisms that cause degradation of pesticides. These are: (1) microbiological processes in soils and water and (2) metabolism of pesticides that are ingested by organisms as part of their food supply. While both processes are beneficial in the sense that pesticide toxicity is reduced, metabolic processes do cause adverse effects in, for example, fish. Energy used to metabolise pesticides and other xenobiotics (foreign chemicals) is not available for other body functions and can seriously impair growth and reproduction of the organism.

Degradation of Pesticides in Soil

Many pesticides dissipate rapidly in soils. This process is mineralisation and results in the conversion of the pesticide into simpler compounds such H_2O, CO_2 and NH_3. While some of this process is a result of chemical reactions such as hydrolysis and photolysis, microbiological catabolism and metabolism is usually the major route of mineralisation.

Soil micro biota utilise the pesticide as a source of carbon or other nutrients. Some chemicals (for example 2,4-D) are quite rapidly broken down in soil while others are less easily attacked (2,4,5-T). Some chemicals are very persistent and are only slowly broken down (atrazine).

Process of Metabolism

Metabolism of pesticides in animals is an important mechanism by which organisms protect themselves from the toxic effects of xenobiotics (foreign chemicals) in their food supply. In the organism, the chemical is transformed into a less toxic form and either excreted or stored in the organism. Different organs, especially the liver, may be involved, depending on the chemical. Enzymes play an important role in the metabolic process and the presence of certain enzymes, especially "mixed" function oxygenases (MFOs) in liver, is now used as an indicator that the organism has been exposed to foreign chemicals.

PESTICIDE MONITORING IN SURFACE WATER

Monitoring data for pesticides are generally poor in much of the world and especially in developing countries. Key pesticides are included in the monitoring schedule of most Western countries, however, the cost of analysis and the necessity to sample at critical times of the year (linked to periods of pesticide use) often preclude development of an extensive data set. Many developing countries have difficulty carrying out organic chemical analysis due to problems of inadequate facilities, impure reagents and financial constraints. New techniques using immunoassay procedures for presence/absence of specific pesticides may reduce costs and increase reliability. Immunoassay tests are available for triazines, acid amides, carbamates, 2,4-D/phenoxy acid, paraquot and aldrin.

Data on pesticide residues in fish for lipophilic compounds and determination of exposure and/or impact of fish to lipophobic pesticides through liver and/or bile analysis is mainly restricted to research programmes. Hence, it is often difficult to determine the presence, pathways and fate of the range of pesticides that are now used in large parts of the world. In contrast, the ecosystemic impacts from older, organochlorine pesticides such as DDT, became readily apparent and has resulted in the banning of these compounds in many parts of the world for agricultural purposes. Older pesticides, together with other hydrophobic carcinogens such as PAHs and PCBs, are poorly monitored when using water samples. As an example, the range of concentration of suspended solids in rivers is often between 100 and 1000 mg/l except during major runoff events when concentrations can greatly exceed these values. Tropical rivers that are unimpacted by development have very low suspended sediment concentrations, but increasingly these are a rarity due to agricultural expansion and deforestation in tropical countries. As an example, approximately 67% of DDT is transported in association with suspended matter at sediment concentrations as low as 100 mg/l and increases to 93% at 1000 mg/l of suspended sediment.

Given the analytical problems of inadequate detection levels and poor quality control in many laboratories of the developing countries, plus the fact that recovery rates vary from 50-150% for organic compounds, it follows that monitoring data from water samples are usually a poor indication of the level of pesticide pollution for compounds that are primarily associated with the solid phase. The number of NDs in many databases is almost certainly an artifact of the wrong sampling medium (water) and, in some cases, inadequate analytical facilities and procedures.

Clearly, this makes pesticide assessment in water difficult in large parts of the world. Experience suggests that sediment-associated pesticide levels are often much higher than recorded and NDs are often quite misleading. Some water quality agencies now use multi-media (water + sediment + biota) sampling in order to more accurately characterize pesticides in the aquatic environment.

Pesticide monitoring requires highly flexible field and laboratory programmes that can respond to periods of pesticide application, which can sample the most appropriate medium (water, sediment, biota), are able to apply detection levels that have meaning for human health and ecosystem protection and which can discriminate between those pesticides which appear as artifacts of historical use versus those that are in current use.

For pesticides that are highly soluble in water, monitoring must be closely linked to periods of pesticide use. In the United States where there have been major studies of the behaviour of pesticide runoff, the triazines (atrazine and cyanazine) and alachlor (chlorinated acetamide) are amongst the most widely used herbicides.

These are used mainly in the spring Studies by Schottler indicate that 55-80% of the pesticide runoff occurred in the month of June. The significance for monitoring is that many newer and soluble pesticides can only be detected shortly after application; therefore, monitoring programmes that are operated on a monthly or quarterly basis (typical of many countries) are unlikely to be able to quantify the presence or determine the significance of pesticides in surface waters. Pesticides that have limited application are even less likely to be detected in surface waters. The danger lies in the presumption by authorities that ND (non-detectable) values implies that pesticides are absent. It may well only mean that monitoring programmes failed to collect data at the appropriate times or analysed the wrong media.

PESTICIDE MANAGEMENT AND CONTROL

Prediction of water quality impacts of pesticides and related land management practices is an essential element of site-specific control options and for the development of generic approaches for pesticide control. Prediction tools are mainly in the form of models. Also, the key hydrological processes that control infiltration and runoff and erosion and sediment transport, are controlling factors in the movement of pesticides.

The European Experience

The Netherlands National Institute of Public Health and Environmental Protection concluded that "groundwater is threatened by pesticides in all European states. This is obvious both from the available monitoring data and calculations concerning pesticide load, soil sensitivity and leaching. It has been calculated that on 65% of all agricultural land the EC standard for the sum of pesticides (0.5 mg/l) will be exceeded. In approximately 25% of the area this standard will be exceeded by more than 10 times" In recognition of pesticide abuse and of environmental and public health impacts the European countries have adopted a variety of measures that include the following:

- Reduction in use of pesticides (by up to 50% in some countries).
- Bans on certain active ingredients.
- Revised pesticide registration criteria.
- Training and licensing of individuals that apply pesticides.
- Reduction of dose and improved scheduling of pesticide application to more effectively meet crop needs and to reduce preventative spraying.
- Testing and approval of spraying apparatus.
- Limitations on aerial spraying.
- Environmental tax on pesticides.
- Promote the use of mechanical and biological alternatives to pesticides.

Elsewhere, as for example Indonesia, reduction in subsidies has reduced the usage of pesticides and has increased the success of integrated pesticide management programmes.

Pesticide Registration

Pesticide control is mainly carried out by a system of national registration which limits the manufacture and/or sale of pesticide products to those that have been approved. In developed countries, registration is a formal process whereby pesticides are examined, in particular, for mammalian toxicity (cancers, teratogenic and mutagenic effects, etc.) and for a range of potential environmental effects based on the measured or estimated environmental behaviour of the product based on its physicochemical properties. Most developing countries have limited capability to carry out their own tests on pesticides and tend to adopt regulatory criteria from the developed world. As our knowledge of the effects of pesticides in the environment accumulates, it has become apparent that many of the older pesticides have inadequate registration criteria and are being re-evaluated. As a consequence, the environmental effects of many of the older pesticides are now recognised as so serious that they are banned from production or sale in many countries.

A dilemma in many developing countries is that many older pesticides (e.g., DDT) are cheap and effective. Moreover, regulations are often not enforced with the result that many pesticides that are, in fact, banned, are openly sold and used in agricultural practice. The dichotomy between actual pesticide use and official policy on pesticide use is, in many countries, far apart. Regulatory control in many countries is ineffective without a variety of other measures, such as education, incentives, etc. The extent to which these are effective in developed versus developing countries depends very much on:

- The ability of government to effectively regulate and levy taxes.
- On the ability or readiness of the farming community to understand and act upon educational programmes.

The fundamental dilemma remains one of accommodating local and short term gain by the farmer (and manufacturer and/or importer) by application of an environmentally dangerous pesticides, with societal good by the act of limiting or banning its use. There is now such concern over environmental and, in some instances, human health effects of excessive use and abuse of pesticides within many governments of the need to include a programme of pesticide reduction as part of a larger strategy of sustainable agriculture. In 1992, Denmark, the Netherlands and Sweden were the first of the 24 member states of the OECD to embark upon such a programme. The Netherlands is the world's second largest exporter of agricultural produce after the United States. In contrast, wood preservatives in the forest sector account for 70% of Swedish pesticide use with agriculture using only 30%. As noted above, the lack of baseline data on pesticides in surface waters of OECD countries, is a constraint in establishing baseline values against which performance of the pesticide reduction programme can be measured.

Index

P

R

S

T

U

V

W